Agro Enterprises for Empowering Farm Women

About the Author

Dr. T. Rathakrishnan is working as Director, Students Welfare, Tamil Nadu Agricultural University, Coimbatore. He has initiated the dual degree programme at under–graduate level with Dalhousie University, NOVA SCOTIA, Agricultural College Campus, CANADA. He has introduced the BEC Programme for the benefit of students.

He is an extension scientist specialized in Communication and Agricultural Extension Management. Also he has worked in Regional Research stations, KVKs and Agricultural Colleges at Tirunelveli, Madurai and Coimbatore in various capacities. He has rendered his 33 years of service in teaching, research and extension.

He has won many medals and awards viz., William Kirubabai medal (Gold) for Best Student in Ph.D. (Agri. extension) in 1993, T. Nataraj Rolling Shield (MASU) for Best Extension Researcher in 2005, Shiksha Ratan Puraskar (IIFS) and Best Extension Worker Award (TNAU) in 2006, and Best Citizen Award (International Publishing House, New Delhi) in 2007.

He has published 14 books, and more than 140 research papers, popular articles and abstracts in seminars/symposiums at National and International level. He has visited IFPRI, Washington (Collaborative project on TOT) and University of Reading UK (PGDRD&E program). Advances and challenges in Rural Development and Extension, Traditional Agricultural Practices, Communication Techniques in Farm Extension, Participatory Technology Development are the text books recently published by him.

Currently, he is operating the International Collaborative Research project on 'Precision Farming for Enhancing the Livelihood Security of Farmers' funded by SHASTRI Indo Canadian Institute, New Delhi as Indian Coordinator with Dalhousie University, Canada.

Mrs. S.R. Padma is a Ph.D., in Agricultural Extension at Tamil Nadu Agricultural University. She has the additional degrees viz., MHRM (Master of Human Resource Management), Dip. T&D (Diploma in Training and Development).

She has won many medals and awards viz., Sri Jagajivan Ram Award, Prof. G. Rangaswamy Medal, and Late Janab T.P M. Mohammed Hameeth of Vandavasi Award for the best M.Sc.(Agri) student in Agricultural Extension in the year 2001.

She has served as Senior Research Fellow in Agricultural Extension, Coimbatore in the Scheme 'Action research on designing effective communication strategies in the system of broad based agricultural extension' and also in 'TN-IAMWARM' project in TNAU for the period of three years. She has performed as Technical Officer in Godrej Pest Management Services, and Technical and Marketing Coordinator in Carees Industries, a micronutrient company. She has participated in 14 National and International Seminars/Workshops/Conferences, and has presented research papers also. She has Best Paper and Best Presentation Awards to her credit. She has published more than 30 research papers, popular articles, research abstracts in various reputed Journals and Seminars. She has been closely associated with Professional Societies of her discipline as a member at National and International level. She is one of the authors of the book on Participatory Technology Development.

Currently, she is associated with Dalhousie University, Canada as Student Researcher in the International Collaborative Research Project on 'Precision farming for enhancing the livelihood security of farmers' funded by SHASTRI Indo Canadian Institute, New Delhi.

Agro Enterprises for Empowering Farm Women

Dr. T. Rathakrishnan
Mrs. S.R. Padma

SCIENTIFIC
PUBLISHERS (INDIA)

Published by:

Scientific Publishers (India)
5 A, New Pali Road, P.O. Box 91
Jodhpur 342 001 (India)

Branch Office
Scientific Publishers (India)
4806/24, Ansari Road, Daryaganj
New Delhi - 110 002 (India)

E-mail: info@scientificpub.com
Website: www.scientificpub.com

ISBN: 978-81-7233-855-8

Typeset by Rajesh Ojha
Printed in India

Preface

Development of entrepreneurship among women is a major step to increase women participation in the process of economic development. It will speed up economic growth, and provide employment opportunities for women resulting in improving the economic independence. Provision of economic opportunities for women can also improve the social, educational and health status of women and their families.

The hidden entrepreneurial potentials of women have gradually been changing with the growing sensitivity to the role women play and their economic status in the society. The participation of women in economic activities is necessary not only from a human resource point of view but also is essential even from the objective of raising the status of women in the society. The economic status of women is now accepted as an indicator of a society's stage of development and therefore, it becomes imperative for the government to frame policies for development of entrepreneurship among women.

The role of farm women entrepreneur in economic development is also being recognized and steps are being taken to promote agro enterprises. Resurgence of agro entrepreneurship is the need of the hour emphasizing on educating and acquiring skill through trainings, spreading awareness and consciousness amongst farm women to outshine in the enterprise field, making them realize their strengths. Women entrepreneurship must be moulded properly with entrepreneurial traits and skills to meet the changes in trends, challenges in global markets and also be competent enough to sustain and strive for excellence in the entrepreneurial arena.

Though there are several agro enterprises available to farm women, selection of enterprises depends on the availability of raw materials, scope for sustaining and establishing market network. This book gives the broad idea about the importance of such enterprises in national and global level, the critical areas in which farm women require training and the strategies to combat the constraints in establishing agro enterprises.

This book will serve as a guide for the farm women engaging in agro enterprises and will also be highly useful as a reference material to the scholars and scientists who have been involved in the process of entrepreneurial development.

AUTHORS

Contents

1.	**Introduction**	**1-7**
	1.1. Women entrepreneurs	1
	1.2. Why women to become entrepreneurs?	1
	1.3. Reasons for slow progress of women entrepreneurs in india	2
	1.4. Need for developing entrepreneurship among farm women	4
	1.5. Significance of the study	6
	1.6. Operationalization of the terms used	7
2.	**Theoretical orientation**	**8-40**
	2.1. Entrepreneur–entrepreneurship - definitions	8
	2.2. Socio–psychological determinants of farm women	12
	2.3. Characteristics of entrepreneurs	18
	2.4. Concepts, meanings and definitions of empowerment	24
	2.5. Training needs of farm women	28
	2.6. Association of socio--psychological determinants with training needs	30
	2.7. Impediments encountered by entrepreneurs	31
	2.8. Strategies to overcome impediments	36
3.	**Research methodology**	**41-54**
	3.1. Selection of the study area	41
	3.2. Selection of respondents	42
	3.3. Selection, operationalization and measurement of variables	42
	3.4. Method of investigation	51
	3.5. Statistical tools used	52
4.	**Findings and discussion**	**55-119**
	4.1. Socio–psychological determinants of farm women	55
	4.2. Training need assessment	60

4.3. Training needs of the respondents in selected agro enterprises for entrepreneurial development 73

4.4. Relationship and contribution of independent variables towards training need in selected enterprises 98

4.5. Path analysis 102

4.6. Challenges of farm women in establishing agro enterprises 107

5. Summary and conclusion 120-130

5.1. Methodology 120

5.2. Salient findings 121

5.3. Implications of the study 123

5.4. Suggestions for future area of research 125

5.5. Suggestions for the growth of women entrepreneurs 125

5.6. Initiatives taken by the government 127

5.7. Conclusion 129

References 131-143

Index 144

Chapter - 1

INTRODUCTION

To awaken the people it is the women who must be awakened. Once she is on the move, the family moves, the village moves, the nation moves.

— Jawaharlal Nehru

1.1. WOMEN ENTREPRENEURS

Women Entrepreneurs may be defined as the women or a group of women who initiate, organize and operate a business enterprise. The Government of India has defined women entrepreneurship as 'an enterprise owned and controlled by women having a minimum financial interest of 51.00 per cent of the capital and giving at least 51.00 per cent of the employment generated in the enterprise to women.' Women entrepreneurs engaged in business due to push and pull factors, encourage them to have an independent occupation and to stand on their own feet. A sense towards independent decision–making on their life and career is the motivational factor behind this urge. Saddled with household chores and domestic responsibilities, women want to get independence. Under the influence of these factors, the women entrepreneurs choose a profession as a challenge and as an urge to do something new. Such a situation is described as pull factors. While in push factors women get engaged in business activities due to family compulsion and the responsibilities that are thrust upon them.

1.2. WHY WOMEN TO BECOME ENTREPRENEURS?

In India, the entry of women into business is traced out as an extension of their kitchen activities; mainly 3P's i.e., Pickle, Powder and Pappad. But with the spread of education and passage of time, women have started shifting from 3P's to modern 3E's i.e., Energy, Electronics and Engineering. Skill, knowledge and adaptability in business are the main reasons for women to enter into business ventures. Women Entrepreneur is a woman who accepts challenging roles to meet her personal needs and to become economically independent. A strong desire to do something positive is an inbuilt quality of entrepreneurial woman, who is capable of contributing values in both family and social life. With the advent of media, women are aware of their own traits, rights and also the work situations. The challenges and opportunities provided to the women of digital era are growing so rapidly that the job seekers are turning into job creators. Many women start a business due to some traumatic event such as divorce, discrimination due to pregnancy or the corporate glass ceiling, the health of a family member, or economic reasons such as a lay–off. But a new talent pool of women entrepreneurs is forming today, as more women are opting to leave corporate world to chart their own destinies.

1.3. REASONS FOR SLOW PROGRESS OF WOMEN ENTREPRENEURS IN INDIA

The impediments felt by women entrepreneurs restrict the expansion of women entrepreneurship. The major barriers encountered by women entrepreneurs are as follows:

- The greatest deterrent to women entrepreneurs is that they are women. A kind of patriarchal - male dominant social order is a blockade in their way towards business success. Male members think it is a big risk financing the ventures run by women.
- In a male–dominated society, women are not treated equal to men. This old and outdated social outlook stops women from entering in the field of entrepreneurship. Women are always under a social pressure, which restrains them to prosper and achieve success in the field of entrepreneurship.
- Women entrepreneurs have to face a stiff competition from the men entrepreneurs who easily involve in the promotion and development area and carry out easy marketing of their products. Such a competition ultimately results in the liquidation of women entrepreneurs.
- Indian women give more emphasis to family ties and relationships. Married women have to make a fine balance between business and family. These family obligations also bar them from becoming successful entrepreneurs. Only few women are able to manage both home and business efficiently, devoting enough time to perform all their responsibilities as per priority.

- Absence of proper support, cooperation and back–up for women by their own family members force them to drop the idea of excelling in the enterprise field. The people from the outside world such as financial institutions discourage women entrepreneurs on the belief that at any time they can leave their business and become housewives again.
- Women in India lead a protected life. They are less educated, economically unstable and also they are not self–sufficient. This reduces their ability to bear risks and uncertainties involved in a business unit,
- Lack of self–confidence, will–power, strong mental outlook and optimistic attitude amongst women creates a fear from committing mistakes while doing their piece of work. The educational level and family background of husbands also influences women participation in the field of enterprise.
- Many women attend the Entrepreneurial Development Programme without an entrepreneurial view in mind. Women who are imparted training by various institutes must be verified on account of aptitude through the tests, interviews etc.
- High production cost of some business operations adversely affects the development of entrepreneurs. The entrepreneurs are not able to handle the factors such as installations of new machines during expansion of the productive capacity. There is no exception for women entrepreneurs also.
- The sincere efforts taken by some of the institutions in the financial sectors towards the empowerment of women entrepreneurs may not reach the rural and backward areas due to lack of awareness about the financial assistance in the form of incentives, loans, schemes etc.
- Low level education coupled with low level confidence is the cause for lowest degree of achievement motivation among the farm women engaged in business.

Apart from the above discussed problems, there may be other serious problems faced by women entrepreneurs such as improper infrastructural facilities, high cost of production, modern business outlook and low needs of enterprise. Also on an average, women tend to start business about ten years later than men. Motherhood, lack of management experience and traditional socialization has all been cited as reasons for delayed entry into entrepreneurial careers. These issues and the possible ways to overcome them are discussed in the subsequent sections.

Twentieth century witnessed the entry of women into professions due to expansion of educational facilities for them. Women are no longer confined to the four walls of their home. More and more women are seeking employment outside their families. The sheer economic need and the gradual spread of education were largely responsible for this change. In recent years, women are making their mark even in the field of business and industry. The government and non–government agencies are giving greater attention to promote self–employment among women.

These concerted efforts supported by higher level of education among women, political awakening, urbanization, legal safe–guards and social changes have facilitated the emergence of women entrepreneurship in India.

But the efforts to develop entrepreneurship among women are focussed more on urban women. It may not be an exaggeration to say that rural women have been totally ignored in the development process. This is not a good sign of development. As per 1991 census there were 90 million female workforce in India, of which 77.85 million (86.5 per cent) were employed in rural areas. Hence, the development of women is a reality only when the rural women are developed.

1.4. NEED FOR DEVELOPING ENTREPRENEURSHIP AMONG FARM WOMEN

In the developing countries, it is observed that unemployment and underemployment are greater for women than for men. Men move out of agricultural employment more easily than women. Moreover, increasing the efficiency of agriculture has been widely shown to have adverse effects, on the employment of women, due to the realignment of tasks in the sexual division or labour. Only now it is beginning to be realized that plans for development must have components aimed specifically at women, if women are to benefit from them. In this context, it has been suggested that the upgrading of women's productive activities by the provision of income–generating employment outside the home may break the cycle of rural poverty and high fertility characteristics of economically–stagnant rural economics.

Efforts to promote small enterprise development in rural areas have until now been directed towards men only. Women are equally endowed with motivation and managerial capabilities in starting and running small enterprises located in their own homes or communities.

Promoting and developing such entrepreneurship among rural women will require initiative, encouragement and support from the development agencies.

It must be recognized that as long as total income in most of the rural communities of Asia is inadequate to maintain acceptable standards of living for the family as a whole, attempts to improve the position, specifically of women, are not likely to proceed far beyond the realm of good intentions. This calls our attention to the urgent need for developing entrepreneurship among farm women.

Entrepreneurship is today regarded as an important tool of development in all parts of the world. Therefore, an effort at involving women in the mainstream of development must include their training as entrepreneurs. Such programmes, apart from concentrating on entrepreneurial awareness, training and credit reach for women, must also concentrate upon bringing the new technology within the reach of women and acquaint them with the new avenues of entrepreneurship. Entrepreneurship development requires a strong support system consisting of credit, marketing, infrastructure and training.

Training is a major catalytic force for augmenting human productivity in all spheres of development. Training of farm women can play a crucial role in providing necessary technical knowledge, attitude and skill required by them for taking up self–employment ventures.

Each development activity especially agriculture requires a back–up of effective training for human resource development for accelerating farm production. Training farm women is an useful tool to impart knowledge and develop new skills required for adoption of latest technology and build up scientific attitude among them. It is a costly investment and must be planned, budgeted and evaluated with care. And the trainings are being imparted by government agencies, non–government agencies, institutions, and government projects.

To improve the knowledge and skill of small and marginal farm women, there are many programmes of Government of India having the primary objective of identifying productive farm women and imparting the knowledge about crop–oriented technologies to them through women Agricultural Officers.

Apart from this, to get additional income and upgrade the standard of living of the farm women, many training programmes related to the entrepreneurial activities are being imparted on fruits and vegetables preservation and processing, mushroom production, layer production, sericulture, kitchen gardening, apiculture and livestock management.

Then too, the farm women are not generally very confident and they are bogged down by the concerns of traditions and customs that may undermine their confidence in taking up the role of an entrepreneur.

Therefore, it is essential to identify the area in which they really lack so that training can be imparted in the required areas to farm women to become good entrepreneurs.

Thus, keeping the above points in mind, the present study was aimed to concentrate on assessing the training needs of farm women for entrepreneurial development. The specific objectives are as follows:

- To assess the socio–psychological determinants of farm women
- To ascertain the training needs of the farm women in selected agro enterprises for entrepreneurial development
- To study the relationship between personal characteristics and the training need
- To assess the areas that need to be focussed in establishing the enterprises for their sustainability

1.5. SIGNIFICANCE OF THE STUDY

Entrepreneurs play a key role in the economic development of a country. In our country where human resources are found to be a plenty, we can identify individuals in all segments of population who have the requisite entrepreneurial skills. The employment of women is expected to function as a significant prong in strategies to increase productivity and alleviate poverty.

Innumerable trainings were given to rural women through various organizations under different rural development programmes. Even after this, majority of the trained farm women are yet to take up self–employment. This shows that something is lacking in them for starting an enterprise. So, assessing the training needs of the farm women to develop the entrepreneurship is essential. The present study was undertaken with this in view. The result will give a lead for the areas in which the training is to be imparted and still to be developed. This study will serve for organising skill–oriented training programmes, which in turn ensure the establishment of new enterprises.

The study was conducted in one agricultural division of the State, which represents a small unit only. Therefore, the findings of the study have limitations and are applicable to only those areas with similar agro–climatic conditions and to those with similar standard of living of the farm women.

1.6 OPERATIONALIZATION OF THE TERMS USED

- **Entrepreneur**

A woman entrepreneur is a farm woman who takes an initiative, establishes and maintains an economic activity or enterprise or selected enterprises, namely mushroom production, layer production and value addition of fruits and vegetables.

- **Entrepreneurship**

Entrepreneurship is the ability to discover an investment opportunity and to organize a money–making enterprise contributing to real economic growth.

- **Value addition of fruits and vegetables**

It is operationalized as the conversion of the fresh produce into their value–added products such as squashes, jam, jelly, candies, bars, pickles and the alike from common fruits.

Chapter - 2

THEORETICAL ORIENTATION

Familiarization with earlier pertinent studies was considered essential to comprehend the theoretical framework and to develop the research methodology of the present study. An attempt was, therefore, made to review the studies related, directly and indirectly, to the present task.

2.1. ENTREPRENEUR–ENTREPRENEURSHIP - DEFINITIONS

Entrepreneur

Let us start with defining the terms entrepreneur and entrepreneurship.

The word 'entrepreneur' appeared in the French language long before there was any general concept of an entrepreneurial function.

Schumpeter (1934) described the entrepreneur as 'the innovator who introduces something new into an economy' and stressed the fact that the entrepreneur is the decision maker in a particular cultural context, who commands a range of behaviours that exploit these opportunities.

Stressman (1968) described entrepreneurs as 'information seekers'.

Heredero (1979) described agricultural entrepreneur as 'a person who produces changes, which directly or indirectly lead to higher agricultural output'.

Mohiuddin (1987) described the concept of rural women entrepreneur as 'rural women who undertook to organize and run enterprise'.

Samiuddin (1987) defined an entrepreneur as 'one who detects and evaluates the new situation in his environment and directs the making of such adjustments in the economic system as he deemed necessary'.

Setty (1990) described an entrepreneur as 'the one who would start an industry or take to a new enterprise, which is a deviation from his traditional family occupation or profession'.

To exploit the human resource to their optimum potential is one of the greatest challenges faced by innumerable business organizations in the contemporary economic world, where progress is quantified and consumed as an entity. (Brush 1992)

Chatterjee (1992) defined an entrepreneur as 'one who creates something new, undertakes risks, organises production and handles the economic uncertainty'. He termed entrepreneurship as the mission and entrepreneur as the missionary.

In Oxford English Dictionary (1993) entrepreneur is defined as 'one who undertakes an enterprise, especially contractor acting as intermediary between capital and labour'.

Sarmah and Singh (1994) stated that an entrepreneur is 'one who can transform raw materials into goods and services, who can effectively utilize physical and financial resources for creating wealth, income and employment, who can innovate new products, standardise or upgrade existing products for creating new markets and new customers'.

Desai (1995) reported that entrepreneur is 'one who can see possibilities in a given situation where others see none and has the patience to work out the idea into a scheme to which financial support can be provided'.

Education of women have, no doubt, given them immense confidence and encouragement to serve and discover new business avenues. However, the ground realities differ to a great extent as far as geographical boundaries are concerned. Globally, women are considered as weaker gender, physically and emotionally. Therefore, prospects open for them to develop into business professionals is an area still quite unexplored and needs attention. (Wennekers 1999)

Entrepreneurship is not just confined to any one gender rather now due to multi–faceted economic pressures women have turned up and realized that the survival of their families and their own potential lies only in working side by side with men. (Marlow 2002)

We can define an entrepreneur as 'one who undertakes a commercial enterprise and who is an organizational creator and innovator'. (Gartner *et al.,* 2004)

In any activity, we cannot eliminate the characteristics of human behaviour. Winn (2005) established a definition on such characteristics as 'One who prospects for or exploits opportunities and who has a tenacity to face challenges'.

Entrepreneurs 'Seek out and identify potentially profitable economic opportunities [and as such are] agents of growth'. (Wharton and Brunetto 2007)

Singh and Belwal (2008) stated that, "The word entrepreneur is derived from the French verb enterprendre that means to undertake".

While defining entrepreneur, some writers have employed economic indicators like growth and profit, or incorporated entrepreneurship as an effective tool in reducing poverty.

Timmons (2009) stated that, "Entrepreneurial behaviour is about utilising resources beyond the immediate scope of the entrepreneur and his or her venture".

The authors came across a gender–specific definition that specifically appealed to them due to its emphasis on female entrepreneurship. 'Female entrepreneurs are defined as those who use their knowledge and resources to develop or create new business opportunities, who are actively involved in managing their businesses, and own at least 50.00 per cent of the business and have been in operation for longer than a year" (Wharton and Brunetto, 2009)

Entrepreneurship

Stevenson (1983) stated that, "Entrepreneurship is the pursuit of opportunity without regard to the resources currently controlled".

Entrepreneurship has been globally felt as a developmental and progressive idea for business world. (Scott 1986)

Reddy (1989) stated that, "Entrepreneurship is a mental urge to take risk in the face of uncertainties and intuition, and capacities of seeing things in a way, which afterwards prove to be true".

Anjaneyaswamy (1992) opined that entrepreneurship in essence lies in the shedding of inhibiting value system and in imbibing of new values relevant to the emerging realities of the environment.

Entrepreneurship is the ability to coordinate and organize, manage and maintain and reap the best out of even the worst situations reported in the study of Vijayalakshmi (1992).

In the opinion of Sarmah and Singh (1994) entrepreneurship is essentially a function, creativity and behaviour manifestation of a person for shifting resources from areas of low productivity to higher productivity.

According to Desai (1995) entrepreneurship is the propensity of mind to take calculated risks with confidence to achieve a pre–determined business.

This women empowerment is exactly the real barrier, which society does not acknowledge and therefore majority of woman potential remains untapped, which can be a powerful resource in economic growth of every economy. (Helms 1997)

Certain writers portray entrepreneurship as a human behaviour linked to personal achievement "Entrepreneurial behaviour is often driven by diverse reasons including the desire for personal accomplishment" (OECD 1998)

For a woman entrepreneurship is not a new concept but certainly makes the woman hesitant to be among start ups due to societal and community trends and pressures. Many other factors influence women to keep them away from being potential entrepreneurs especially in developing countries. (Hisrich 1999)

Entrepreneurship is an idea or vision, which a woman holds to explore and optimize and could help her create new jobs and economic empowerment among her fellow beings.

Across the globe, generations of women from very different backgrounds show very encouraging sign of entrepreneur spirit. It is to be hoped that governments at all levels will work to provide an environment in which this spirit may flourish. (Delmar 2000)

Although entrepreneurship is usually regarded as an activity, some definitions concentrate more on its operational and functional characteristics.

According to Shane and Venkataraman (2000), 'Entrepreneurship concerns the environment conditioning opportunity, the process of discovering opportunity, the evaluation and exploitation of opportunity, and the individual decision–makers who do these things'.

Entrepreneurship is considered as vital ingredient for not only globalization but at the same time for creating diverse opportunities for future potential performers. (Mitra 2002)

Entrepreneurship by definition implies being in control of one's own life and activities. (Gundry, Yoseph and Posig 2002)

Over 200 million women are employed across all industrial sectors, with half of this number is in developing countries. (Global Entrepreneurial Monitor (GEM) report 2007)

Developed nations have provided gender equality platforms to keep the woman workforce motivated to be entrepreneurs. However, women still face financial insecurity, business insurance, human resource challenges and resistance from their male counterparts to be successful woman entrepreneurs.

2.2. SOCIO–PSYCHOLOGICAL DETERMINANTS OF FARM WOMEN

Age

Perumal (1994) reported that majority of the participants (70.83 per cent) belonged to young category, followed by middle aged ones (29.17 per cent).

According to Parvathi (1995), Murugesan (1996), Karthikeyan, Premavathi (1997), the majority of farm women were young.

Zamir Ahmed (1998) observed that majority (58.00 per cent) of farm women trainers were found to be young aged.

According to GEM annual report (2007) in the low and middle–income country groups, women are most likely to be early stage entrepreneurs between the ages of 25–34 yr and to become established entrepreneurs between the ages of 35–44 yr.

In high income countries, the age window for women entrepreneurial activity broadens, with early stage entrepreneurial activity most likely among women aged 25–44 yr and established business ownership most likely among women 35–54 yr old regardless of gender or country group.

Educational status

Shanmugavadivu (1992) stated that 42.50, 17.50, 17.00, 15.50 and 5.50 per cent of farm women had primary, illiterate, secondary, middle and higher secondary levels of education respectively.

Jayasree (1993) reported that majority of participants had medium level of education (59.01 per cent) followed by 28.80 per cent with low educational status and 12.19 per cent with high educational status.

Uma (1994) observed that one–third (35.83 per cent) of the farm women possessed middle level of education followed by illiterates (27.50 per cent), primary education (20.83 per cent), secondary education (10.84 per cent) and collegiate education (5.00 per cent). There was none found to be functionally literate among the respondents.

Muthammal (1997) found that more than one–third (35.83 per cent) had middle education followed by illiterates (25.00 per cent) and secondary school education (23.33 per cent). Primary school education was found among 14.16 per cent of the respondents.

On an average, women entrepreneurs in high–income countries are better educated than those in low and middle–income countries. In high–income countries, more than half of women entrepreneurs have secondary degrees, and more than one–quarter have graduate degrees. In low and middle–income countries, the percentage of women as early stage entrepreneurs and established business owners with less than a secondary degree is 34.1 per cent and 40.4 per cent, respectively. (GEM Report 2007)

Occupational status

Devasena (1992) studied the occupational status of women and found that only 30.00 per cent of the respondents had agriculture as primary occupation and 40.00 per cent had farming as secondary occupation. The remaining one–third of the respondents desired to work in matchbox factories and fire work industries.

Desigavinayagam (1994) revealed that nearly half of the Krishi Vigyan Kendra (KVK) trainees (53.33 per cent) had farming alone as their occupation. Nearly one–fifth of the respondents had farming and labour as their occupation.

Perumal (1994) reported that more than half of the trainees (57.50 per cent) had farming and labour as their occupation and less than half of them (42.50 per cent) had farming alone as their occupation. None of the trainees had farming and business as their occupation.

Annual income

Kalivaradhan (1990) and Ponmani (1993) reported that more than half of the rural women respondents (60.00 per cent) had low level of income.

Prameela (1992) stated that half of the respondents (50.66 per cent) belonged to medium–income group, followed by 42.67 per cent under high–income group.

Chockalingam (1994) reported that 38.33 per cent of the farm women respondents were in the medium–income group followed by 35.00 and 26.67 per cent in the low and high–income groups respectively.

Uma (1994) stated that a vast majority (91.67 per cent) of the rural women respondents belonged to low–income group.

Suresh (1997) found that nearly three–fourths of the respondents (72.50 per cent) belonged to low–income group.

Information source utilization

Helen *et al.* (1992) reported that more than one–fourth of adult members used input agencies (36.40 per cent) and Assistant Agricultural Officers (30.10 per cent) for seeking information. Cent per cent of the farmers utilized neighbours as their non–institutional source of information. All the farm women had utilized neighbours as their source of information and majority of the farm women (97.90 per cent) got information from family members. Farm broadcast was listened by 84.40 per cent and more than 50.00 per cent of the farmers got information from newspapers.

Prameela (1992) observed that majority (75.33 and 68.33 per cent) of the farm women had contacted Assistant Agricultural Officers and Agricultural Officers in the 'often' and 'sometimes' frequency category respectively. Neighbours, friends and relatives were the often utilized channels by 78.67, 77.33 and 62.00 per cent respectively. Radio was utilized by 51.34 per cent of the farm women in the 'sometimes' frequency category.

According to Mary *et al.* (1993) the sources of information to the youths of small farm families were their parents, relatives, neighbours, friends and radio whereas to youths belonging to big farm families the sources were their parents, relatives, extension agents and television.

Nirmaladevi (1997) revealed that majority of the respondents considered the Agricultural Officers and Assistant Agricultural Officers as their institutional sources of information.

Payal (1999) reported that majority of the respondents contacted Agricultural Demonstrator, followed by Agricultural Supervisor, Agricultural Inspector and Assistant Agricultural Officer. They were the institutional sources available at village level. She also reported that among the non–institutional sources, majority of the respondents contacted friends, relatives, progressive farmers and agricultural leaders. She also added that radio, newspaper and television were the important media sources utilized by the farmers.

Innovativeness

Ravichandran (1992) found out that majority of the farm women respondents (60.00 per cent) had medium level of innovativeness followed by low (28.00 per cent) and high (12.00 per cent) levels of innovativeness.

Chockalingam (1994) stated that 55.00 per cent of farm women had high level of innovativeness followed by medium (27.50 per cent) and low (17.50 per cent) levels.

Sivarevathi (1996) observed that 45.00 per cent of farm women possessed medium level of innovativeness followed by high (30.83 per cent) and low levels of innovativeness (24.17 per cent).

Muthammal (1997) reported that two–third (75.00 per cent) of the farm women had medium level of innovativeness followed by low level (25.00 per cent).

Premavathi (1997) stated that majority (57 per cent) of the farm women trainees had a high degree of innovativeness.

As per the remarks of Noorjehan (1999) most of the respondents had medium level of innovation proneness (54.00 per cent) followed by low (34.00 per cent) and high (12.00 per cent) levels.

Self-confidence

Sheela (2000) reported that an overwhelming majority of (80.84 per cent) farm women possessed medium to high level of self–confidence. Only 19.16 per cent possessed low level of self–confidence.

Decision making

Rexlin and Subramaniyan (1990) found that 26.00 per cent of small farm women and 33.00 per cent of big farm women were consulting their husbands for decision making on farm practices.

Ravichandran (1992) reported that the extent of participation of farm women in making self–decisions on farm production activities was 27.00 per cent and 35.00 per cent in farm investment activities. The extent of their participation in making joint decisions of farm production activities was 73.00 per cent, while they influenced 65.00 per cent investment activity.

Mohanthy (1994) observed that as far as the purchase of farm equipments and animals were concerned, majority of the decisions were taken either by the husband alone (35.25 per cent) or by husband and wife jointly (36.25 per cent).

Monica and Talukdar (1997) reported that majority (63.34 per cent) of the women entrepreneurs possessed high decision making ability followed by medium and low level in equal percentage (18.33 per cent in each category).

Baby Kumari (1998) quoted that majority of the farm women respondents took independent (self) decisions.

Seetha Lakshmi (1999) found that majority of the farm women (93.33 per cent) made joint decisions by having consultation with family members for farm and home activities. Five per cent of the respondents had independent decisions and 1.67 per cent had joint decision by having consultation with other than family members.

Level of aspiration

Arularasan (1992) reported that an overwhelming majority of respondents (72.50 per cent) belonged to low aspiration category.

Pandian (1999) stated that 44.17 per cent of farm women had low level of aspiration followed by 38.33 per cent in medium and 17.50 per cent in high level categories.

Economic motivation

Ravichandran (1992) reported that most of the rural women respondents had high levels of economic motivation followed by medium (7.00 per cent) level of economic motivation. None possessed low level of economic motivation.

Savithiri (1992) found out that 61.33 per cent farm women in dry tracts were seen with high level of economic motivation followed by low and medium levels (24.67 per cent and 14.00 per cent), respectively.

Chockalingam (1994) opined that nearly half of the farm women (48.33 per cent) had medium level of economic motivation followed by low (42.50 per cent) and high (9.17 per cent) levels.

Uma (1994) found out that rural women had high economic motivation since 47.50 per cent of the women trainees had high level of economic motivation followed by middle level (34.17 per cent).

Sivarevathi (1996) observed that a little less than half of the farm women (44.17 per cent) possessed low level economic motivation followed by medium (29.17 per cent) and high (26.66 per cent) levels.

Muthammal (1997) revealed that more than half (55.83 per cent) of the farm women possessed low levels of economic motivation followed by high (25.00 per cent) and medium (19.17 per cent) levels.

Attitude towards self-employment

Monica and Talukdar (1997) observed that 45.00 per cent of women entrepreneurs had favourable attitude towards self–

employment followed by 41.66 per cent and 13.33 per cent with less favourable and more favourable attitudes respectively.

Trainings

Study of Ponmani (1993) revealed that majority of the rural women (90.00 per cent) did not attend any training prior to the participation in the studied training programme and the rest (10.00 per cent) were found to have attended one training.

Sivarevathy (1996) found out that majority of farm women participants of development programme (88.33 per cent) did not attend any training. About one–tenth (9.17 per cent) were found to have attended one training and only 2.50 per cent attended two trainings. None had attended more than two trainings.

Muthammal (1997) observed that most of the farm women (92.50 per cent) did not attend any training. Less number (7.50 per cent) of respondents were found to have attended training.

Arunadevi (1998) opined that 36.00 per cent of the trainees had attended two trainings, 32.80 per cent attended more than two trainings, while 31.20 per cent attended only one training.

Seetha Lakshmi (1999) pointed out that majority of the farm women (75.00 per cent) participated in one to two training programmes and around 20.00 per cent of them did not participate in any of the training programmes.

Perceived effectiveness of training

Ray (1993) pointed that 55.00 per cent of the trainees opined that the content of the training course was good. Overwhelming majority (95.00 per cent) felt that it was practical oriented to a limited extent.

Further, majority (65.00 per cent) of the trainees felt that expectations were fairly fulfilled. All the trainees opined that objectives of the training were suited to their requirements. In respect of overall assessment of the quality of the course 65.00 per cent and 35.00 per cent of the trainees assessed it to be good and very good respectively.

According to Chandrakandan *et al.* (1994) majority of the farmers received the effectiveness of the training at high level.

As per Vijayakhader (1994), after adequate training, the trained farm women had shown a much more positive attitude towards mushroom cultivation.

Majority of the farm women trainees (72.00 per cent) were satisfied with the applicability of the subject matter taught to them with special reference to the cultivation of crops as pointed out by Sundararajan (1995).

As opined by Murugesan (1996) majority of the trainees perceived the effectiveness of the training at medium level.

Karthikeyan (1997) inferred that majority of the trainees perceived that the information given in the training programme was more adequate, highly practical, more useful, easily understandable, presented in order, well–explained and well–stressed.

Attitude towards training

Perumal (1994) reported that prevalence of high level of attitude was observed with 45.83 per cent of trainees followed by 33.33 per cent with medium and 20.84 per cent with low level towards the training programme.

Uma (1994) revealed that little more than one–third of the Training Rural Youth for Self Employment (TRYSEM) trained farm women (35.00 per cent) had medium level of attitude towards training programme and another one–third respondents (33.33 per cent) had high level of attitude, followed by 31.67 per cent respondents with low level of attitude.

Poovannan (1995) reported that almost an equal proportion of the trainees had less favourable (41.00 per cent) and more favourable (39.00 per cent) attitude towards training. One–fifth of them (20.00 per cent) possessed favourable attitude.

Ravichandran (1997) revealed that 83.57 per cent of trainees had more favourable attitude towards training, followed by 12.86 per cent and 3.57 per cent with less favourable and favourable attitudes, respectively.

Muthammal (1997) stated that more than half (67.50 per cent) of the farm women had favourable attitude towards training and little less than one–third (31.67 per cent) of the farm women had more favourable attitude towards training.

2.3. CHARACTERISTICS OF ENTREPRENEURS

The characteristics of an entrepreneur farmer identified by Epstein (1967) are as follows:

- High education
- Contacts with various people

- Innovator
- Enterprising cultivator, who adopts improved practices in farming

Agarwal (1975) stated that coordination, planning and uncertainty bearing or innovation are the three main descriptions of entrepreneurs.

Ramakrishnan (1979) had enlisted the characteristics of entrepreneurs as, high level of aspiration, managerial competence, self–confidence, leadership qualities, risk–taking ability and independence in thought and action.

Singh (1979) revealed the successful characteristics of tribal entrepreneurs as follows:

- High achievement motive, affiliation motive and power motive
- Innovative and cosmopolite
- Tactful and independent

Sridaran (1981) observed that all the entrepreneurs had high level of income after taking up the sericulture enterprise.

According to Singh and Sehgal (1986), characteristics of entrepreneur are: high degree of achievement motivation, capability to take calculated risk, a positive image of one's own abilities and achievements, realistic, self–assessment of one's strength and weakness, problem solving approach, initiative and drive, ability to think independently, resistance and ability to cope with setbacks, imagination and creative ability and change proneness.

Timmons (1986) listed 14 characteristics of entrepreneurs, which are as follows:

1. Drive and energy 2. Self–confidence 3. Long term involvement 4. Money as measure 5. Persistent problem solving 6. Goal setting 7. Moderate risk taking 8. Dealing with failure 9. Use of feedback 10. Taking initiative and seeking personal responsibility 11. Tactful use of resources 12. Competing with self–imposed standard 13. Internal locus of control 14. Tolerance for ambiguity and uncertainty

Indeed women face multidimensional challenges like external financing and credibility as business owners and managers.

Additional barriers include socialization networks and practices, family roles and possible lack of business contacts. Disadvantages experienced by women in all these areas are widely believed to result in higher failure rates and lower growth rates for women–owned businesses (Bowen and Hisrich 1986).

Adult women representing a large group of the overall population in each country is a readily–available pool of potential entrepreneurs that can generate economic growth activity to improve their overall economies. However, the countries differ in stages of development, demographic patterns and labour conditions (Stevenson 1986).

Dixit (1988) stated that Indian entrepreneurs both at rural and urban centres with guts, skills and ambition did exist but qualities like motivation, sense of commitment, morality were invariably absent in small entrepreneurs.

Positive projection of entrepreneurial activity in a country should be an integral element of any government agency in order to boost economic prosperity, which is subsequently a way out of poverty and economic crisis.

Maricle and Birkenholz (1988) had entitled the characteristics of entrepreneurs as, a strong internal locus of control, high tolerance of ambiguity, high independence, high energy level, ability to face uncertainty, a futuristic outlook, persistent, creative, willingness to take risks, low tolerance to frustration, tension and dissatisfaction with the system.

Women, by nature, preferably choose to work in service sector like education, social services and household services such as domestic work. But at the same time the start–up motivations vary from industry to industry. (Birley 1989)

Perumal (1990) reported the profile of women entrepreneur as - mostly belonged to young age, had secondary level of education, had high level of experience, had encouraging level of social participation, use of mass media to a greater extent and majority of the farm women had attended specialized and skill–oriented trainings.

While analyzing the entrepreneurial behaviour of sugarcane growers, Pandya (1996) observed that more than half of the respondents were having education up to secondary level, had medium level of social participation, medium farm power, medium income, material possession and extension contact.

Motivational factors influencing woman to join the leading group of entrepreneurs are of two types - one is entrepreneurship by choice, and the other is entrepreneurship by necessity (Schumpeter 2000).

Women become entrepreneurs by choice due to the following factors:

- To materialize their idea into a capital
- For their empowerment and freedom
- To prove their worth among their male family members
- To establish their own rules for their work
- To overcome the deficiencies they faced during their job experience
- To fulfil a long term standing desire to own their own company
- Working for someone else did not appeal to them

Women become entrepreneurs by necessity due to the following factors:

- To improve the quality of life of their children
- To share the family economic burden
- To adjust and manage household and business life successfully on their own terms
- Due to the death or sickness of their husband/father/brother

Other than these motivational factors some secondary motivations also exist among women entrepreneurs.

- Women want fair treatment and improved compensation
- Prior industry and work experience plays a very important role in determining their start–ups' success.
- Women believe that prior experience is crucial
- Many women believe that a track record is particularly valuable under these conditions because it demonstrates their confidence.

Certain social factors also prevail:

- Socio–economic factors
- Money is not the sole objective
- Building their own self–confidence
- Want to contribute something positive to the community
- Bring out their creativity and turn it into an innovation
- Materialize their ideas
- Economic independence
- Establishing their own creative idea
- Establishing their own identity

- Achievement of excellence
- Developing risk taking ability
- Motivation
- Equal status in society
- Greater freedom and mobility

The growth issue is particularly important in high–income countries where women are increasingly entering into technology–based business ventures. As compared to male–owned business, the female–owned businesses tend to be smaller and grow more slowly. Generally, smaller sizes and slower growth rates are perceived as problems. (Shane and Venkataraman 2000)

For a woman, entrepreneurship is a way out of economic disparity and towards empowerment. Her entrepreneurship contributes positively in various dimensions and aspects in economic development and job creation (Gundry and Welsch 2001).

As a result, women in low–income countries have strong incentives to create their own work environments that are compatible with their child rearing and household responsibilities at the same time (GEM report 2004).

This tendency is more strengthened by the typically lower degree of involvement in child care by fathers in poorer countries. Consequently, women preferably look for home–based business opportunities (Brush, Carter and Gatewood 2006).

Women entrepreneurship is expanding around the world. Studies have shown that women–owned businesses comprise between one–quarter and one–third of the businesses in formal economy and are likely to play even greater role in informal sectors (GEM report 2007).

More female entrepreneurship exists in developing countries where traditionally higher birth rates prevail. The law and customs of marriage have higher rate of influence whether a woman chooses to be an entrepreneur (Monolova 2008).

Generally, the attitudes and employment practices in low–income countries are less likely to provide child care, health care and training opportunities to women as employees. At the same time, employers are less likely to provide a comfortable workplace and a set of adequate work practices that allow a woman to both perform satisfactorily on the job and care appropriately for her children (Muhammad 2008).

Factors influencing new business venture preferences among women entrepreneurs

Embedding is particularly vital for female entrepreneurship since woman entrepreneur behaviour is deeply moulded by family structure and social ties in both high and low–income countries. Different cultural context may produce a different policy outcome (Star and Yudkin 1996).

The preferences of the women who perceive themselves as having necessary skills and knowledge to be business owners is higher as compared to women who are lacking in these skills. As a result, greater emphasis lies on individual awareness and perception processes, which consequently point out the importance of local social norms (Littunen 2000).

It depends on the market demand and supply side, which determines the entrepreneurial environment within which women receive incentives to take appropriate decisions about entrepreneurship (Orhan and Scott 2001).

Transparency, the rule of law and well–defined property rights promote productive entrepreneurship in all times and places. Policies regarding equal opportunities need to be framed because women entrepreneurs do not act or perform in seclusion from their society and community (GEM report 2004).

The likelihood of starting new ventures in business depends significantly on the aptitudes, skills and knowledge level of women entrepreneurs (Smith 2005).

Across the globe, generations of women from very different backgrounds show very encouraging signs of entrepreneur spirit. It is hoped that governments at all levels will work to provide an environment in which this spirit may flourish.

Particular traits of women entrepreneurs

Women entrepreneurs are potentially more confident and perceive the existing unexploited opportunities than women who are non–entrepreneurs. Women perceive their paid jobs as a platform, which enable them to move towards their entrepreneurial activity as their future career objective.

Individuals involved in entrepreneurial activity, irrespective of their age of business, clearly differ in their own skills and are more likely to know other entrepreneurs as well. Women who opt to be entrepreneurs tend to be optimistic about their own knowledge and capabilities. Entrepreneurs exhibit higher level of self–confidence and belief in their own abilities. This self–confidence consequently may help increase their awareness to recognize their opportunities and thus the creation of more new firms. (Charboneau 1981)

Research has shown that age, work, status, education, income, social ties and perceptions are still significant socio–economic factors in a person's decision to start a business. (Birley 1989)

The GEM 2007 data implies that fear of failure is in fact highly correlated to entrepreneurial activity, which is the risk–taking attribute that primarily prevails among all entrepreneurs. Fear of failure withholds one–third of women population to become entrepreneurs.

2.4. CONCEPTS, MEANINGS AND DEFINITIONS OF EMPOWERMENT

Sudharani *et al.* (2000) defined empowerment as the process of challenging existing power relations and gaining greater control over the sources of power. Empowerment is a process of awareness and capacity building leading to greater participation to greater decision–making power and control to take transformative action.

Hemantha (2001) reported that the role of university in empowering women has comprehensively given the parameters of empowerment of women as follows:

- Enhance self–esteem and self–confidence in women
- Build a positive image of women by recognizing their contribution to the society and economy
- Develop in them an ability to think critically, faster decision–making action through collective process
- Enable women to make informed choices in areas such as education, employment and health especially reproductive health
- Ensure equal participation in development process
- Provide information, knowledge and skill for economic independence
- Enhance access to legal literacy and information related to their rights and entitlements in the society with a view to enhance their participation in all areas

Mayoux (2001) suggested that empowerment is a process of internal change, or power within, augmentation of capabilities, or power to, and collective mobilization of women, and when possible men, or power with, to the purpose of questioning and changing the subordination connected with gender, or power over.

Nirmala and Ravishankar (2001) defined that empowerment can be interpreted as creating power, increasing power or enabling power to be exercised. It is not power over any one, but power for women to determine own action, power to have the authority to make decision and to have access to resources.

According to Selvin *et al.* (2001), women need to be empowered educationally, socially, psychologically, economically, technologically and politically. The first step towards the capacity building of women is allowing women to avail their basic human rights to education. Worthless, negative attitude should be changed. Economic empowerment is the pre–requisite for women's holistic development. For that they should be technologically empowered as the process of technology has not given due recognition to women. And they are to be politically empowered, which we have attained to some extent, with the passing of the 73rd Amendment Act. Fortunately the development programme like Tamil Nadu Women in Agriculture (TANWA), Development of Women and Children in Rural areas (DWCRA) and Self Help Groups (SHGs) have contributed positively in empowering women in social, psychological, economic and technical aspects to some extent.

World Bank (2001) report confirms that societies that discriminate on the basis of gender pay the cost of greater poverty, slower economic growth, weaker governance and a lower living standard of their people. The World Bank also identifies four key elements of empowerment to draft institutional reforms: access to information, inclusion and participation, accountability and local organisational capacity.

Desmarai (2002) reported that empowerment means social justice, gender and ethnic equality, economic equality and environmental sustainability.

Malhotra *et al.* (2002) constructed a list of the most commonly used dimensions of women empowerment, drawing from the frameworks developed by various authors in different fields of social sciences. Allowing for overlap, these frameworks suggest that women empowerment needs to occur along multiple dimensions including economic, socio–cultural, familial or interpersonal, legal, political, and psychological. Since these dimensions cover a broad range of factors, women may be empowered within one of these sub–domains. They give the example of socio–cultural dimension, which covers a range of empowerment sub–domains, from marriage systems to norms regarding physical mobility of women, to non–familial social support systems and networks available to women.

Reddy (2002) stated the term empowerment as a redistribution of social power and control of resources in favour of women.

Whirter (2002) revealed the term empowerment as - become aware of power dynamics, develop the skill and capacity, control without infringing upon the rights of others, support the empowerment of others in the community.

Abdul Kalam (2003) spelt out four pronged ways for empowerment- Education, employment generation, economic empowerment, and value–based technology generation.

According to Krishna (2003) empowerment means increasing the capacity of individuals or groups to make effective development and life choices, and to transform these choices into desired actions and outcomes. It is by nature a process and/or outcome.

Sonia Gandhi (2003) stated that the empowerment of women called for a massive national effort to equip them with skills and expertise, apart from power and resources, to fulfil their constitutional obligations.

Swaminathan (2003) explained that empowerment means reaching unreached in knowledge, skill and technological empowerment of women.

World Bank (2003) defines empowerment as the process of increasing the capacity of individuals or groups to make choices and to transform those choices into desired actions and outcomes. Central to this process is actions, which both build individual and collective assets, and improves the efficiency and fairness of the organizational and institutional context which govern the use of these assets.

Amaresh (2004) focused that the term women empowerment means to empower women in the weaker sections and those belonging to the minorities.

Ananadavalli (2004) focused her views on empowerment of women as - knowledge on science and technology was essential for women empowerment and gender sensitive research should be enhanced to identify their roles and participation.

Gonslaves (2004) stated that empowerment means providing employment guarantee for rural folk for at least 100 work days.

Kabeer (2001) defines empowerment as the expansion in people's ability to make strategic life choices in a context where this ability was previously denied to them. It is important to differentiate strategic life choices from other choices as strategic life choices help to frame other, second order and less consequential choices, which may be important for the quality of one's life but do not constitute its defining parameters.

Empowerment is a process of transition from a state of powerless to a state of relative control over one's life, destiny and environment. It is a transition from a passive situation to a more active situation of control. According to him, it is a process through which people learn to take on socially valuable roles, to exercise social skills, to exert interpersonal influence, to develop commitment, to take responsibility and to acquire political efficacy (Sudan 2004).

Empowerment is an ongoing and dynamic process, which enhances women's (and any other marginalized and alienated groups) abilities to change the structures and ideologies that keep them subordinate. The empowerment process may be broken down in three dimensions - economic, social and political, which reinforce each other. While the economic aspects would include increasing women's access to and command over tangible and intangible resources such as wealth, property, employment, knowledge and information, social aspect would include changing the existing discriminatory ideology and culture, which determine the environment for women's existence. Finally, political process must increase women's presence and influence in the power structure (Meena and Puspa 2005).

Empowerment is defined as a person's capacity to make effective choices, that is, as the capacity to transform choices into desired activities and outcomes. The extent or degree to which a person is empowered is influenced by personal agency (the capacity to make purposive choice) and opportunity structure (the institutional context in which choice is made). Asset endowments are used as indicators of agency. These assets may be psychological, informational, organisational, material, social, financial or human. Opportunity structure is measured by the presence and operation of formal and informal institutions including the laws, regulatory frameworks, and norms governing behaviour (Ruth and Nina 2005).

The great attention and the need for the empowerment of women is not surprising considering that gender-inequality is regarded to be one of the greatest and most important disparities in the world in general and in India specifically (Kumar 2006).

According to Mayoux (2006), microfinance programmes, through their contribution to women's ability to earn an income, may initiate virtuous spirals of economic empowerment, increased well–being for women and their families and wider social and political empowerment.

However, in her analysis of the evidence of gender impact, she finds that women empowerment is not an automatic consequence of women's access to savings and credit or group formation. Furthermore, she stated that there is a conflict between three different paradigms, the feminist empowerment paradigm, the poverty alleviation paradigm and the financial self–sustainability paradigm. In her analysis, the latter two paradigms have been dominant, which means that requirements for sustainability or proven development contribution have led to an evaporation of explicit strategies for women empowerment and a progressive narrowing of the definitions of women empowerment.

From the above analysis of literature, it is interpreted that empowerment is a process of awareness and capacity building, which leads to a greater participation and greater decision making power. It also revealed that empowerment of women enhances their innate vitality, which should enable the women or group of women to realize their full identity and power in all spheres of life and also equal partner in development process.

2.5. TRAINING NEEDS OF FARM WOMEN

Whenever a training programme is planned, a basic question arises - Do we or do we not have a need for training? The answer to this question pre–supposes that training should be based on assessment of training needs.

According to Johnson (1967), assessment of training need is important for the following reasons:

- To determine and meet specific growth needs
- To organize the training based on valid, present or emerging needs so that the money, time and efforts are not wasted
- To facilitate need–based and quality training

For organizing quality training programme, the training needs of farm women should be assessed and then they should be deputed for attending the training programmes. This process will facilitate proper utilization of high amount of expenditure incurred on training. The trainees will also take keen interest in training activities and the trainers will also be motivated to conduct qualitative and need–based training.

According to Gupta (1982), training needs of farmers were perceived high in order of importance in respect of crop husbandry followed by poultry and dairy. The next in order was fruits and vegetables preservation and bee keeping. Rabbit rearing and mushroom cultivation were viewed as somewhat important for training. Lastly farmers had shown their least concern for training in piggery and fishery.

Singh *et al.* (1989) stated that dairy farmers needed training mostly in clean milk production and housing of animal (59.5 per cent) followed by veterinary care (46.8 per cent) and in feeding practices (11.2 per cent).

According to the study conducted by Narmatha (1990), the training need aspects of women poultry farmers are, medication, delousing, disinfection, culling of unproductive sick birds, purchase of feed and construction of shed.

Ponnusamy (1990) identified that most of the farm women (93.00 per cent) needed training in animal husbandry followed by bee keeping (58.00 per cent) and mushroom cultivation (53.4 per cent).

Helan *et al.* (1992) reported that 49.00 per cent of farm women needed training in dairy farming, followed by poultry farming (33.3 per cent) and mushroom cultivation (20.8 per cent).

Pushpa (1992) observed that majority of the poultry farm women (88.2 per cent) required training in disease management followed by selection of birds (57.6 per cent) and feed preparation (52.9 per cent).

Lakshmi Raju *et al.* (1994) stated that dairy farm women needed more training in selection of animals, health care, management of animals, feed preparation, balanced feeding, calf management, marketing aspects followed by cattle shed management, proper milking, dehorning of calf, and preventive vaccination.

Usha and Annapurna (1996) reported that kitchen gardening, food preservation and processing, mushroom cultivation were the major enterprises in which the farm women required more training.

According to Mathiyalagan Peru and Subramanian (1998) majority of the poultry farm women (88.00 per cent) needed more training in disease prevention and control followed by feed ingredients (73.43 per cent), and feed mixing and brooder management (60.54 per cent). They needed very less training in structure of poultry housing and marketing. And they added that nearly three–fourths of the poultry farmers preferred to have the peripatetic training in their own village as the venue of training and they preferred March–April as the suitable season for training.

Sailaja and Reddy (1999) observed that the farm women needed more training in mushroom cultivation, sericulture, apiculture enterprises and they needed less training in fruits and vegetables preservation, goat farming and poultry farming.

Zamir Ahmed *et al.* (2000) found that a vast majority of the farm women (85.00 per cent) required more training in kitchen gardening, mushroom cultivation and layer farming and more than half of the farm women (55.00 per cent) opted KVK as the venue for the training and 15.00 per cent of farm women preferred farmers field as the venue for conducting the training.

And also they observed that most of the farm women (80.00 per cent) preferred four days and above as duration of training and 20.00 per cent have an open choice of any number of days as duration of training.

2.6. ASSOCIATION OF SOCIO--PSYCHOLOGICAL DETERMINANTS WITH TRAINING NEEDS

On the basis of review of research findings of various research workers the relationship of characteristics with training need are given in Table 1.

Table 1. Relationship of characteristics with training need

S. N.	Variable	Author	Year	Relationship with training need
1.	Age	Thiruvirchezian	1985	Positive and significant
		Thamariselvi	1989	Negative and significant
		Alagesan	1990	Negative and significant
		Rose	1991	Negative and significant
		Desigavinayagam	1994	Non–significant
2.	Educational status	Veeraputhiran	1988	Positive and significant
		Thamariselvi	1989	Negative and significant
		Alagesan	1990	Non–significant
		Alagesan *et al.*	1990	Positive and significant
		Sekar *et al.*	1990	Negative and significant
		Rose	1991	Negative and significant
		Thangachamy	1993	Non–significant
		Desigavinayagam	1994	Positive and significant
		Muthammal	1997	Non–significant
3.	Occupational status	Alagesan	1990	Non–significant
		Alagesan *et al.*	1990	Non–significant

		Sekar *et al.*	1990	Non–significant
		Rose	1991	Negative and significant
		Desigavinayagam	1994	Non–significant
4.	Innovativeness	Paramasivam	1988	Positive and significant
		Velumani	1988	Positive and significant
		Sutha	1989	Positive and significant
		Dharmalingam	1990	Positive and significant
		Desigavinayagam	1994	Non–significant
5.	Economic motivation	Thamariselvi	1989	Positive and significant
		Sekar *et al.*	1990	Positive and significant
		Rose	1991	Negative and significant
		Alagesan	1990	Non–significant
		Desigavinayagam	1994	Non–significant
		Veeraputhiran	1988	Positive and significant

2.7. IMPEDIMENTS ENCOUNTERED BY ENTREPRENEURS

Mathai (1978) enlisted the four risks faced by entrepreneurs.

1. Technical risks

This includes the risk of not knowing the technical details and therefore, not being able to solve them.

2. Economical / financial risks

This includes the risk of market fluctuation and changes with regard to raw materials and finished products.

3. Social risks

This includes the risks in the development of new relationships within the village and outside. The entrepreneur also often has to deny himself the pleasures attached to family and social life in order to escape from the risk.

4. Environmental risks

This includes the risks as a result of environmental changes in the villager's work as an outcome of the new activity, and the risks inherent in dealing with unfamiliar people, cultures, rate of change, system etc.

Ayyadurai (1980) reported that lack of finance is the main constraint as reported by majority of respondents. Lack of knowledge about the institutions that help in poultry enterprises, dislike among the family members, disinterest towards poultry, risk and uncertainty etc., were as in the order of sequence.

Sethy (1982) quoted the following difficulties in poultry enterprise:

- Non–availability of high quality chicks from the government
- High risk in price of poultry feed
- Lack of facilities for marketing of eggs and birds
- Absence of veterinary infrastructure facilities
- Lack of extension education to the farmers and supply of inputs
- Local coordination between banks and various extension agencies

Mohiuddin (1987) reported that rural women entrepreneurs faced the difficulties as men entrepreneurs did, such as provision of raw material, loans and marketing. Added to these were certain discriminations; sometimes in the positive direction by providing with more help and guidance or sometimes in the negative direction by rejecting women simply out of the stereotype prejudices.

Some of the problems of poultry industry in South India were poor quality of chicks, shortage of feed, incidence of diseases, inadequacy of extension services and disorganized market system as stated by Panda (1989).

The poultry farms in Punjab faced some problems as stated by Taneja (1990). They were finance, poor quality chicks, feed, disease management, marketing, technical guidance and difficulty in maintenance of the poultry unit.

According to Arularasan (1992), the major constraints encountered by the youths in setting up self–employment ventures were financial, marketing and seasonal.

Nabi and Mohanty (1992) enumerated the problems of entrepreneurship development as absence of skill and ability of people, backdrop of centuries of natural calamities of flood, drought and famine, lack of education, lack of infrastructural facilities, non–availability of raw materials, transport and communication facilities, cheap and regular supply of power, trained labour, lack of conducive socio–economic environment.

Sreen and Sreen (1992) reported that the most common problems encountered by women in dairy enterprises related to lack of adequate medical care facilities, unavailability of technical personnel, poor feed practices, lack of knowledge and lack of ready access to medicines.

According to Taori *et al.* (1994), problem areas in bee keeping were identified as:

- Bee diseases like TSBV, acquarine and European fowl blood.
- Low difference between purchase price and sales price of honey.
- Commercial production and sale of value added products of honey with ginger, lemon fruits and biscuits etc.
- Standard purchasing and production of standard/graded honey, lack of arrangement of timely collection, proper storage, processing of honey with moisture reduction.
- Multiplication of bee colonies, insurance coverage of bee keeping.
- Extension of consultancy and servicing, exports of honey, commercial production and utilization of pollen, bee venom, royal jelly and wax etc.

Thara (1996) observed that insufficient technical guidance, inconsistency in decision making on crop rotation, site selection, inadequate input availability were the major constraints involved in kitchen gardening.

According to Thenamutha (1996) non–remunerative egg prices, (100.00 per cent), high cost of feed (90.00 per cent), presence of IBD disease (79.17 per cent), high capital investment (79.17 per cent), maintenance of health of birds (68.33 per cent) involvement of risk and uncertainties (60.00 per cent) were the problems faced by poultry entrepreneurs.

Venkataramaiah and Manjula (1996) observed that non–availability of credit, inadequate availability of inputs, lack of marketing facilities for disposing the products, lack of proper transport facilities for reaching the work place were the major problems faced by the women entrepreneurs.

Jeyalakshimi *et al.* (1997) reported that the most common problems encountered by the women entrepreneurs in mushroom cultivation were - marketing problems, lack of financial assistance, non–availability of spawn and preservation of produce. They also indicated that the constraints faced by the women entrepreneurs in fruits and vegetables preservation were - marketing problems, high cost of fruits, lack of financial assistance, difficulty in getting license and lack of demand for the product.

According to Ravichandran (1997), lack of availability of plant protection chemicals and fertilizers for mulberry cultivation, lack of knowledge about the disinfectants for silk worm rearing and fluctuation in market price for cocoon were the major problems faced by sericulturists.

Anand (1998) classified the major problems that hinder rural women empowerment as:

- Individual and family related problems
- Illiteracy
- Inadequate knowledge on matters relating to health and family life
- Early marriage, unwanted pregnancies, unsanitary living conditions
- Burden of hard and long hours of gruelling work
- Lack of decision making at home
- Physical abuse and abandonment at home
- Community level problems
- Socio–cultural and religious resistance to development of women
- Inhibition of women to participate in important activities
- Lack of mobilization and organized effort from activists
- Lack of community related programmes
- Socio–political problems
- Women's unawareness of laws and welfare schemes
- Existing laws protecting women are not enforced
- Justice is slow and costly
- Media projects a distorted image of women
- Economic problems
- Economic dependency on men

Chidambaram and Thenmozhi (1998) enlisted the following constraints encountered by the women entrepreneurs:

- Personal constraints
- Educational constraints

- Socio–physico sphere system constraints
- Marketing constraints
- Labour and raw material constraints
- Technological constraints
- Financial constraints
- Support system constraints

According to Suseela Mary (1998), lack of suitable control measures for TSBV disease, unremunerative price for honey, not getting immediate money for honey from the cooperative societies, lack of sufficient financial help from government and lending institutions and high cost of sugar were the constraints expressed by apiculturists. She added that usage of pesticides in the migration centres and occurrence of pests and diseases on the honey bee forage crop were also the major constraints.

Govindappa (1999) found that the women entrepreneurs faced social risks and technical risks.

Sehrawat *et al.* (1999) revealed that very serious marketing problems encountered by entrepreneurs were cheaper (or) superior competitive substitutes, inadequate supply of export information and high cost of packaging were found as very serious problems, while unfamiliarity with export activities and identification of mega markets or potential markets for exports, lack of marketing intelligence, quality specification and finishing, lack of sophisticating technology of packaging for export, competition from already established and larger units, cumbersome procedure of import and export were also found as serious problems.

Amudha and Veerabhadraiah (2000) reported that major problems experienced by farm women in poultry farming were lack of knowledge about improved practices and disease control measures, higher feed cost and low egg prices, lack of storage facilities for eggs, low egg consumption, lack of export facilities and their dependency on middle men to get poultry feed and marketing their eggs.

Neelima and Swaroop (2000) pointed out that lack of family support, lack of access to peer support, low level of education, lack of general exposure, strong fear of failure and ridicule, lack of recognition, lack of property and assets, and non–availability of collateral and block of access to capital were the problems faced by the women entrepreneurs.

2.8. STRATEGIES TO OVERCOME IMPEDIMENTS

Efforts are being taken for equality of opportunity in all spheres to the Indian women, laws guaranteed equal rights of participation in political process, and equal opportunities and rights in education and employment were enacted. But unfortunately, the government–sponsored development activities have benefited only a small section of women i.e., the urban middle class women. Women sector occupies nearly 45.00 per cent of the Indian population. At this juncture, effective steps are needed to provide entrepreneurial awareness, orientation and skill development programs to women. The role of women entrepreneurs in economic development is also being recognized and steps are being taken to promote women entrepreneurship. Resurgence of entrepreneurship is the need of the hour emphasizing on educating women strata of population, spreading awareness and consciousness amongst women to outshine in the enterprise field, making them realize their strengths, and importantly their position in the society and the great contribution they can make for their industry as well as the entire economy. Women entrepreneurship must be moulded properly with entrepreneurial traits and skills to meet the changes in trends, challenges in the global markets and also be competent enough to sustain and strive for excellence in the entrepreneurial arena.

Nadkarni (1988) pointed out that the spirit of entrepreneurship could be nurtured to some extent by an appropriate pattern of education and training programme.

Himachalam (1990) gave the suggestions for entrepreneurship development as - frequent market surveys, suitable organisational arrangements for disseminating information about appropriate technology, full assistance in meeting the financial requirements, follow–up support, sponsoring of Technical Consultancy Organisations (TCOs) for providing necessary facilities and guidance to experts so as to enable the entrepreneurs to take their help in the formulation and implementation of the projects, technical training to the entrepreneurs etc.

Phukan (1992) quoted that the National Industrial Policy of 1990 stressed on four major points on the issue of entrepreneurship promotion in the country. They were:

- Support to first generation entrepreneurs through training and encouragement of association of industries in the process
- Dovetailing EDP into vocational and other technical curricula
- Special development effort for women entrepreneurs

Beegam and Sarngadharan (1994) suggested that entrepreneurship programmes among the women might be looked at from two angles. One was to create the environment needed for healthy and sourced entrepreneurship, and the other was to organize more and more rational motivational campaigns for the women entrepreneurs. To a large majority of Indian women, the highest motivation for work was indeed sheer economic necessity.

According to Venkataramaiah and Manjula (1996), provision of more amount of credit by government for running the entrepreneurs, provision of general stores, shops, depots for the supply of essential commodities, and giving training to women to update their knowledge, were the suggestions offered by women entrepreneurs to solve the problems faced by them.

Amudha and Veerabhadraiah (2000) reported the following suggestions to solve the problems of farm women in poultry farming:

- Training programmes should be organized to update the knowledge of farm women about improved poultry practices and disease control measures.
- Veterinary services should be extended to all the poultry farmers by the government to reduce the expenditure.
- Regulatory marketing facilities should be provided for marketing of eggs to fetch remunerative prices.
- Egg consumption should be popularised among the public.
- Cold storage facilities should be provided in each district to avoid the spoilage of eggs.

Sheela (2009) gave the following measures in order to improve the living standard of rural women:

- Provision of skill–oriented training for self–employment
- Alternative livelihood options during the lean season
- Subsidies to start small–scale enterprises
- Educating women to know their rights
- Development of a good marketing network
- Health education for women

Conceptual Model

In common parlance, the entrepreneurial development process involves the general status of the environment, opportunities available and the capacity of the individual to capitalize and nurture further. The various components involved in the sequence of the process are depicted in Fig. 1.

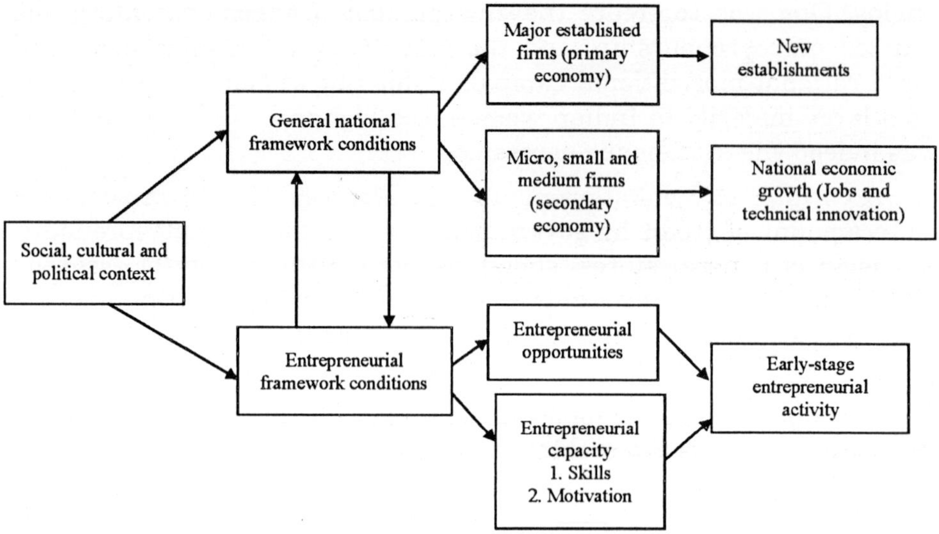

Fig. 1. Various components involved in entrepreneurial development process (*Source:* GEM Report 2007)

Stages of Empowerment

According to Anand (1998), empowerment can be visualized at two levels:

1. Individual empowerment
2. Group / Community empowerment that occurs through three stages:
    - **Stage 1** - Dependence: At this stage there is a hierarchical dependence on others for everything.
    - **Stage 2** - Independence: There is development of an ability to take individual or collective decisions, without the intervention of any intermediary or external force and hence this stage is characterized by growth of confidence, rebelliousness and the need to ascertain the limits of one's empowerment.
    - **Stage 3** - At this stage there is increased consciousness among individuals that no man is an island and therefore, learning on the basis of a new kind of dependence based on equality, cooperation and trust in others occur. Sengupta (1998) proposed the need for certain skills to be developed for transforming from the

stage of dependency to the stage of interdependency, viz., develop skills to take actual responsibility, share responsibility, develop effective networks for information and influence, help others in the community or group to acquire personal skills they need to become effective and enhance personal and group creativity.

As a capacity building process, viewing empowerment from the point of the individual(s) and group or community, several strategies may be employed to improve consciousness among people at grass root levels for improving their living conditions and internal dispositions to initiate actions for enacting social change. The process of social learning is being conceptualized in the context of people's participation and empowerment for initiating change at the grass root level. The social learning process in terms of empowerment of people's participation involves the elements like:

- Interactive and interdependent levels of learning
- Common elements of information collection, information interpretation and goal directed action, which are shared by all the levels
- Salient elements of community learning, which aids learning at different levels

Thus, social learning can work as an important empowering strategy for equipping people, groups and community to develop, psychologically and socially, the critical minimum level of awareness and to join together for action by sharing and using their personal capacity or as group or community members. The characteristics of empowered person are:

- Well trained
- Confident
- Enthusiastic, motivated and committed
- Ability to use natural creativity
- Ability to take responsibility
- Ability to communicate needs, success, problems and ideas
- Ability to work on their own in a group
- Ability to take decisions, individually and also by involving others
- Proud of their work
- Be proud of their group to which they belong

- Developing trust in others and ability to create trust in others for oneself
- Comfortable about questioning the status quo
- Ability to understand the context and consequences of their actions
- Knowledge about how well they are performing
- Empowered outside the work place / community level
- Ability to keep learning and developing

To empower the youth, Government of India has initiated different educational, training and employment schemes.

Chapter 3

RESEARCH METHODOLOGY

"Research design is the plan, structure and strategy of investigation conceived so as to obtain answers to the research questions and to control variance. The strategy includes the methods to be used to gather and analyze the data or in other words, the strategy implies how the research objectives could be reached and how the problems encountered in the research would be tackled".

— Kerlinger, 1973

This chapter delineates the research techniques adopted in the study on "Agro enterprises for empowering farm women".

3.1 SELECTION OF THE STUDY AREA

Selection of district

This study was focused on training needs of farm women for entrepreneurial development. The farm women who had been exposed to entrepreneurial activities and had been given sufficient exposure by the programme operators were considered as the sample for this study.

Madurai district of Tamil Nadu was selected deliberately for investigation because this district is one among those operating more programmes. In this district, Tamil Nadu Women Development Project is under operation since 1995.

Selection of the agricultural division

There are six agricultural divisions in Madurai District viz., Madurai South, Madurai North, Vadipatti, Melur, Thirumangalam and Usilampatti. Among these agricultural divisions farm women exposed to various entrepreneurial avenues were more in Thirumangalam agricultural division and hence, the same was selected.

Selection of blocks

Thirumangalam agricultural division comprises of three blocks *viz.*, Thirumangalam, Kalligudi and Kallipatti. Among the three blocks, Thirumangalam is more urban based and hence was excluded from the study. Kalligudi block was deliberately selected from the other two blocks. This selection was made after consulting the agricultural officers for the following reasons:

- Small units of the enterprises, which had considerable proportion of farm women, were selected for this study
- Easy access to the study area from the institute
- The culture and the dialect of the area were familiar to the researcher

Selection of villages

There are 16 Assistant Agricultural Officers groups in Kalligudi block. Among these 16 Assistant Agricultural Officers groups, 14 Assistant Agricultural Officers groups are covered by entrepreneurial avenues. Each group comprises of 47 villages. In consultation with Agricultural Officers, ten villages were termed as progressive and selected based on the following criteria:

- Active participation of farm women in training programmes
- Adoption of selected enterprises by sufficient number of farm women

3.2. SELECTION OF RESPONDENTS

Farm women were considered as the sampling unit for the study. From each of the selected villages a list of farm women was obtained from the office records maintained by the concerned officials. They were arranged in alphabetical order. Random sampling method was used to select 120 respondents from this list.

3.3. SELECTION, OPERATIONALIZATION AND MEASUREMENT OF VARIABLES

Propositions involving theoretically–defined concepts are not directly testable and therefore, the actual test was made in terms of the concepts which could be operationally defined (Blalock 1960). Variable selection, operationalization and their empirical measurements are discussed below:

3.3.1. Independent Variable

The variable that is antecedent to the dependent variable is termed as an independent variable (Kothari, 1985).

In discussion with extension workers, social scientists and advisory committee members and on perusal of literature, 30 variables were identified. These variables could influence the training needs of farm women for entrepreneurial development. These 30 variables were sent to 50 social scientists of the State Agricultural Universities and Research Stations for judgment. A three–point relevancy continuum (most relevant, relevant and least relevant) was proposed. Out of 50 judges, responses were received from 26 judges.

The 'most relevant' response was given three scores, 'relevant' with two scores and 'least relevant' was given one score. Mean score and coefficient of variation were worked out for all the 30 variables. The variables, which had their average score value greater than the overall mean score and also their coefficient of variance less than the overall coefficient of variation, were selected for the study. By this procedure 15 variables were selected for the study and the instruments used to measure them are given in Table 2.

Table 2. List of selected variables with their measurement and scoring procedures

Variable No	Variable	Measurement and scoring procedure used
I	Independent variables	
X_1	Age	Scoring procedure followed by Venkatesan (1997)
X_2	Educational status	Scale developed by Mansingh (1993)
X_3	Occupational status	Scoring procedure followed by Thangachamy (1993)
X_4	Annual income	Scoring procedure developed for the study
X_5	Information source utilization	Scoring procedure followed by Prameela (1992)
X_6	Innovativeness	Scoring procedure followed by Venkatesan (1997)
X_7	Self-confidence	Scoring procedure followed by Syamkumar (1999)
X_8	Decision making	Scoring procedure followed by Ravichandran (1992)
X_9	Level of aspiration	Scoring procedure followed by Jeyalakshmi (1996)
X_{10}	Economic motivation	Scoring procedure developed by Supe (1969) and followed by Gangadharan (1996)

X_{11}	Attitude towards self–employment	Scoring procedure developed by Pradeep Kumar (1993)
X_{12}	Trainings undergone	Scoring procedure followed by Ponmani (1993)
X_{13}	Perceived effectiveness of training	Scoring procedure followed by Murugesan (1996)
X_{14}	Attitude towards training	Scoring procedure followed by Zamir Ahmed (1998)

Age

Age was operationalized as the number of completed years of the respondent at the time of enquiry and the chronological age was taken as the measure. The completed years of the respondents were taken for analysis. The respondents were classified into three categories according to their age as followed by Venkatesan (1997).

Table 3. Age

Age group	Years
Young	Less than 35 years
Middle	Between 35 and 45 years
Old	More than 45 years

Educational status

It referred to the educational qualification of the respondents. The sub–items were illiterate, functionally literate, primary education, middle education, secondary education and collegiate education. Illiterate was an individual who did not know to read and write, whereas functionally literate could read and write. Primary education referred to formal schooling up to 5th standard. Middle education referred to the education in the school from 6th to 8th standard. Secondary education meant the education from 9th standard up to plus two level. Collegiate education referred to degree or diploma after schooling. The scale developed by Mansingh (1993) was followed.

Table 4. Educational status

Category	Score
Illiterate	1
Functionally literate	2
Primary education	3
Middle education	4

Secondary education	5
Collegiate education	6

Occupational status

Occupational status was operationalized as the profession in which an individual spent much of his or her time, money and resources as a means to gain better livelihood to maintain oneself and his or her family. The scoring procedure followed by Thangachamy (1993) was adopted in this study.

Table 5. Occupational status

Occupation	Score
Farming alone	1
Farming and labour	2
Farming and caste occupation	3
Farming and business	4
Farming and independent profession	5
Farming and service	6

Annual income

Annual income was operationalized as the gross annual income obtained by the respondents from various sources in rupees. The scoring procedure for this variable was developed for this study. One score was allotted for every thousand rupees. The respondents were categorized into three groups such as low, medium and high using cumulative frequency method.

Table 6. Annual income

Category	Income (Rs.)
Low income group	Up to 20,000
Medium income group	20,001–40,000
High income group	More than 40,000

Information source utilization

Information source utilization was referred as the frequency of use of information sources by farm women for getting information. The sources were grouped into personal–cosmopolite, personal–localite and impersonal–cosmopolite. The sources considered for this measurement are given below:

Personal–cosmopolite sources

- Assistant Agricultural Officers
- Agricultural Officers
- Agricultural Development Officers
- Assistant Director of Agriculture
- Agricultural Scientists

Personal–localite sources

- Progressive and contact farmers
- Neighbours and relatives
- Input dealers

Impersonal–cosmopolite sources

- Radio
- TV
- Newspapers and magazines

The scoring procedure followed by Prameela (1992) was adopted for this variable.

Table 7. Information source utilization

Frequency	Score
Often	4
Sometimes	3
Rarely	2
Never	1

The respondents were categorized as low, medium and high based on cumulative frequency method.

Innovativeness

Innovativeness was operationalized as the degree to which a farm woman is earlier in adopting new ideas than the other members of the social system. Innovativeness was measured using the scoring procedure followed by Venkatesan (1997).

The scale consisted of five statements and the responses were collected on a five–point continuum as given below:

Table 8. Innovativeness

Response	Score
Strongly agree	7
Agree	5
Undecided	4
Disagree	3
Strongly disagree	1

The sum of the scores for the five statements was taken as the total score for innovativeness. The maximum score one would get was 35, while the minimum was five. The respondents were categorized as low, medium and high based on their innovativeness by using cumulative frequency method.

Self–confidence

Self–confidence was operationalized as the extent of feeling about one's own powers, abilities and resourcefulness to perform any activity, which the farm women desire to undertake.

This variable was measured using the scoring technique followed by Syamkumar (1999). The scoring technique consisted of eight items of which three were positive (Statement No: 1, 2 and 5) and the rest were negative. The respondents were asked to give their responses in a five–point continuum ranging from, 'strongly agree' to 'strongly disagree'. The scoring procedure was as follows:

Table 9. Self–confidence

Respondents	Strongly agree	Agree	Undec-ided	Disagree	Strongly disagree
For positive statements	5	4	3	2	1
For negative statements	1	2	3	4	5

Self–confidence score for each individual was calculated by summing up the scores of an individual. The possible scores of an individual in this scale ranged from 8–40. The respondents were classified as low, medium and high based on the self-confidence score by using cumulative frequency method.

Decision making

Rogers and Shoemaker (1971) defined decision as the process by which an individual engages in activities, which leads a choice to adopt or reject an innovation.

In this study, decision making pattern was referred as the nature of participation of farm women respondents in making decisions on entrepreneurial activities.

A farm woman may take independent decisions or consult her husband or family members or other than family members to take decisions. The scale developed and adopted by Ravichandran (1992) was used in this study.

Table 10. Decision making

Type of decision	Scores
Independent decision	3
Joint decision with family members or husband	2
Joint decision with other than family members	1

The scores obtained by the individual farm women respondent was taken as the decision making score of the particular individual.

Level of aspiration

Level of aspiration was operationalized as the overall assessment of rural women's concern for wishes and hopes for the future or for the fears and worries about the future in her own world of reality.

Level of aspiration in the present study was measured using the scale followed by Jeyalakshmi (1996). The scale consisted of 12 statements with three alternatives provided for each item. Relative weights of 1, 2, 3 were assigned for these alternatives. The scores obtained for each item were summed up to get the score on level of aspiration. The possible score ranges from 12–45. Using cumulative frequency, the respondents were classified as low, medium and high.

Economic motivation

Economic motivation was operationalized as the profit maximization and the relative value placed on economic ends by a rural women.

The dimension was measured using the economic motivation scale developed by Supe (1969) and followed by Gangadharan (1996). The scale consisted of five statements. The responses were measured on a five–point continuum as follows:

Table 11. Economic motivation

Response	Score
Strongly agree	7
Agree	5
Undecided	4
Disagree	3
Strongly disagree	1

The scores obtained for each statement were summed up to arrive at the individual's score on economic motivation. The score ranged from 5–35. Using cumulative frequency the respondents were categorized as low, medium and high.

Attitude towards self–employment

Attitude towards self–employment is defined as the degree of positive or negative feeling of rural women towards self–employment.

In the present study, attitude towards self–employment was measured using the scale developed by Pradeepkumar (1993). The scale consisted of ten statements. The responses were measured on a five–point continuum as given below:

Table 12. Attitude towards self–employment

Respondents	Strongly agree	Agree	Undecided	Dis-agree	Strongly disagree
For positive statements	7	5	4	3	1
For negative statements	1	3	4	5	7

The scores obtained for each item were summed up to arrive at the individual's score on attitude towards self-employment. The possible score ranges from 70–10. Using cumulative frequency the respondents were classified as low, medium and high.

Trainings undergone

In this study, trainings attended by farm women was operationalized as the number of trainings attended by them at the time of data collection, which was taken as a measure by allotting one score each. The scoring procedure followed by Ponmani (1993) was used in this study. '0' score was given for not attending any training and addition of one score for each training attended. The scoring procedure followed is given below:

Table 13. Trainings undergone

Training	Score
No training	0
One training	1
Two trainings	2
Three trainings	3

The percentage of respondents in each of the above category was worked out.

Perceived effectiveness of training

Perceived effectiveness of training was referred as the perceived opinion of the respondents about the training programme. Their expression on the level of perception was obtained on a three–point continuum given for each statement. The scoring procedure followed by Murugesan (1996) was used.

Table 14. Perceived effectiveness of training

Item	Score
More sufficient	3
Sufficient	2
Less sufficient	1

The scores obtained for each statement by an individual respondent were summed up and the total was the perceived effectiveness score for an individual respondent. Further, based on mean and standard deviation it was classified into low, medium and high categories.

Attitude towards training

Attitude was operationally defined as the degree of positive or negative disposition associated with the individual towards training. The scale developed by Sundarajan (1995) having six statements of which first, fourth and sixth were negative. The farm women responses were obtained on a three–point continuum regarding agreement, undecidedness and disagreement with weightages of 3, 2, and 1 for favourable statements and the reverse order for unfavourable statements. The sum of the scores on individual items represented the attitude score for individual farm women. Finally based on mean and standard deviation they were classified into low, medium and high groups.

3.3.2. Dependent Variable

If one variable depends upon other variable or if it is a consequence of the other variable, it is termed as a dependent variable (Kothari 1985). Training need was the dependent variable taken up for this study.

Training need in this study has been operationalized as the required level of training by the respondents on entrepreneurial development.

In consultation with the scientists of plant pathology, animal husbandry and food science, a list of critical areas of the selected enterprises were identified. Final selection of specific areas was based on the results of pre–test.

Training need of each major subject matter area was assessed over a four–point continuum such as 'Very much needed', 'Needed', 'Less needed' and 'Not needed'. And they were quantified by assigning the scores of 4, 3, 2 and 1, respectively. The responses, thus obtained, have been quantified as percentage.

Table 15. Training need

Item	Score
Very much needed	4
Needed	3
Less needed	2
Not needed	1

3.4. METHOD OF INVESTIGATION

Interview schedule was found to be the most appropriate device to collect data from farm women. Interview schedule was prepared in accordance with the objectives of the study. Before finalizing the schedule, it was pre–tested in a non–sampling area. The inconsistencies noted were rectified by discussing with the members of the advisory committee and thus the schedule was finalized.

Selection of enterprises

The following criteria were laid down in the selection of enterprises for assessing the training needs of farm women for entrepreneurial development:

- The enterprises should be technically viable, economically feasible and socially acceptable.

- It should be easily practicable.
- It should be an enterprise directly applicable to the farm women to generate self–employment for increasing their income.
- The respondents should have adequate exposure in the enterprises selected.

Based on the above mentioned criteria, three enterprises, namely layer production, mushroom production and value addition of fruits and vegetables were selected for the study.

3.5. STATISTICAL TOOLS USED

The following statistical tools were used for the analysis and interpretation of the data:

- Arithmetic mean
- Percentage analysis
- Cumulative frequency
- Correlation coefficient
- Multiple regression analysis
- Path analysis

3.5.1. Arithmetic Mean ($\bar{x}$)

Arithmetic mean is the quotient that results when sum of all items in the series is divided by the number of items (N).

$$\bar{x} = \frac{\sum x}{N}$$

Where,

$\bar{x}$ = mean

Σx = Sum of each individual score

N = Total number of items

3.5.2. Percentage Analysis

Percentage analysis was used in descriptive analysis for making simple comparisons. The frequency of the particular cell was multiplied by 100 and divided by the total number of respondents pertaining to particular cell. Percentages were corrected up to two decimal places.

3.5.3. Cumulative Frequency

This method was suggested by Rao (1983) to categorize the respondents into three groups, namely low, medium and high.

Based on the score values, the number of respondents belonging to each class was determined. The square root frequency was calculated. Then the cumulative frequency was worked out. The final cumulative frequency value was multiplied by 1/3 and 2/3 to find out the two boundaries, namely L_1 and L_2, respectively. The exact values of these boundaries were calculated using the following formula:

$$L_i = K + \left[\frac{L_i - C}{f}\right] n$$

Where,

K = Median between lower limit of the class in which L_i occurs and the upper limit of the previous class.

L_i = Boundary values, namely L_1 and L_2

C = Cumulative square root of frequency up to the classes preceding the class in which L_i lies

N = Interval of the class

f = Square root of frequency in the class in which the median lies

Thc categories were formed as detailed below:

- Below L_1 value - Low
- Between L_1 and L_2 value - Medium
- Above L_2 value - High

3.5.4. Correlation Coefficient

Pearson's product moment correlation coefficient was calculated to find out the degree of association between two variables using the following formula:

$$r_{xy} = \frac{\sum xy - \frac{(\sum x)(\sum y)}{n}}{\sqrt{\left[\sum x^2 - \frac{(\sum x)^2}{n}\right]\left[\sum y^2 - \frac{(\sum y)^2}{n}\right]}}$$

where,

r = Coefficient of correlation between x and y

n = Sample size

x = Sum of scores of independent variable x

y = Sum of scores of dependent variable y

The significance of calculated 'r' values was tested for 5.00 per cent and 1.00 per cent levels of significance.

3.5.5. Multiple Regression Analysis

When a factor is dependent on more than one independent factor, the simple correlation analysis cannot reveal the combined relationship. For this purpose, the multiple regression technique was used to reveal the existence of linear relationship between the dependent and independent variables. Hence, the linear multiple regression was applied as one of the statistical tools for the analysis. It takes the general form of:

$$y = a+b_1X_1+b_2X_2+\ldots.+b_nX_n$$

y = Dependent variable

X_1 to X_n = Independent variables

b_1 to b_n = Regression coefficients

a = Intercept

3.5.6. Path Analysis

Path analysis was performed to estimate the direct effect of all independent variables and their indirect effects through other variables towards the dependent variables.

Chapter - 4

FINDINGS AND DISCUSSION

4.1. SOCIO–PSYCHOLOGICAL DETERMINANTS OF FARM WOMEN

In order to know the background and socio–economic aspects, it is important to analyze the socio–personal characteristics of the farm women respondents. This would present a clear picture of the background of the farm women, which in turn, would help in suggesting appropriate policy implications. Fourteen socio–personal characteristics were studied, and relevant data was collected and analysed. The findings are presented below:

4.1.1. Age

Age is one of the most important factors to be considered in this type of study since it induces a farm woman to choose and adopt a new technology. It may reveal the mental maturity of an individual to take decisions for fulfilling her needs. The data gathered is presented in Table 16.

Table 16. Age categorization of respondents (n=120)

Sr. No.	Category	Number	Per cent
1.	Young (below 35 years)	59	49.17
2.	Middle (between 35 and 45 years)	49	40.83
3.	Old (above 45 years)	12	10.00

From Table 16, it could be inferred that majority (90.00 per cent) of the farm women were young and middle aged. And this finding revealed that young farm women were easily motivated to adopt new technology or to start an enterprise. It is quite natural that the young people usually have positive inclination towards any innovation and they do not have any inhibition in accepting the same. This finding is in line with the findings of Jeyasree (1993) who reported that majority of the trainees were young.

4.1.2. Education

One of the factors, which influence the communication skills of the farm women, is the level of their literacy. Therefore, the literacy level of farm women was investigated from illiterate to collegiate level and the results are presented in Table 17.

Table 17. Educational status of respondents (n=120)

Sr. No.	Category	Number	Per cent
1.	Illiterate	9	7.5
2.	Functionally literate	17	14.16
3.	Primary education	47	39.17
4.	Middle education	47	39.17
5.	Secondary education	-	-
6.	Collegiate education	-	-

The above observation leads to the inference that more than three–fourths (78.34 per cent) of the farm women had education, which ranged from primary to middle level. The farm women with higher level of education viz., secondary school education and college education were absent.

The facilities available for education in the study area were only up to middle school level. In rural areas, girls are normally not permitted to go for education outside the village. These would have been the reasons for most of the respondents to have been educated only up to the middle level.

This finding is in conformity with the report of Jeyasree (1993) and Uma (1994) who stated that majority of the farm woman had education only up to middle level.

4.1.3. Occupational Status

The occupational status of a respondent decides the time available to her for other productive or non–productive activities. Hence, an attempt was made to study the occupational status of the farm women and the results of the analysis are presented in Table 18.

The results reveal that majority of the farm women in the selected villages had farming as their main occupation; also because of the availability of land in rural areas. The results also revealed that only 16.64 per cent of the respondents could further engage themselves into independent profession such as petty shops so as to generate additional income during their free time.

Table 18. Occupational status of the farm women (n=120)

Sr. No.	Category	Number	Per cent
1.	Farming alone	100	83.33
2.	Farming and labour	-	-
3.	Farming and caste occupation	-	-
4.	Farming and business	-	-
5.	Farming and independent profession	20	16.67
6.	Farming and service	-	-

No farm woman was observed in other occupational categories studied viz., farming and labour, farming and caste occupation, farming and business, and farming and service.

This finding is in conformity with that of Uma (1994) who also reported that 83.33 per cent of women respondents were found to have agriculture–based occupation of farming or wage earners or both.

4.1.4. Annual Income

The annual income of the family decides the status of an individual in the social system. It influences the decision of the farm women to invest in income generating on-farm activities and off-farm activities.

The annual income includes income from agriculture, income earned through wages, and income from other enterprises. The distribution of respondents according to their annual income is presented in Table 19.

Table 19. Categorization of respondents based on their annual income (n=120)

Sr. No.	Category	Number	Per cent
1.	Low (up to Rs. 20,000)	54	45.00
2.	Medium (Rs. 20,001-40,000)	53	44.16
3.	High (more than Rs. 40,000)	13	10.84

It could be seen from Table 19 that the respondents were equally distributed in the low (45.00 per cent) and medium income (44.16 per cent) categories. Only 13 respondents were earning more than Rs. 40,000.

Since most of the respondents had farming alone as their occupation, they would have been distributed between low and medium income categories.

This finding is in line with Uma (1994) who also found that a vast majority (91.67 per cent) of the rural women respondents belonged to low–income category.

4.1.5. Information Source Utilization

Unless and otherwise the respondents receive information from all possible sources viz., personal–localite, personal–cosmopolite and impersonal–cosmopolite, their investment on income–generating activities will be restricted. So, it is inevitable for any study related to training need, to know about the respondents' level of utilization of information sources. Pertinent data was gathered and its analysis is presented in Table 20.

Table 20. Information source utilization pattern of the respondents (n=120)

Sr. No.	Category	Number	Per cent
1.	Low	31	25.83
2.	Medium	57	47.50
3.	High	32	26.67

Table 20 reveals that nearly half (47.50 per cent) of the respondents were seen in the medium level, regarding information source utilization. Almost an equal proportion was observed under high (26.67 per cent) and low (25.83 per cent) information source utilization categories.

As the respondents regularly received information on farm technologies through the officials, they did not have the inclination towards other sources and that could be the possible reason for such an outcome.

This result derived the support from the finding of Maheswari (2000).

4.1.6. Innovativeness

Data was collected and analyzed to know about the tendency of an individual farm woman to try and use the new and novel ideas relatively earlier than others in the social system. The results of the analysis are depicted in Table 21.

Table 21. Distribution of respondents according to their degree of innovativeness (n=120)

Sr. No.	Category	Number	Per cent
1.	Low	13	10.83
2.	Medium	86	71.67
3.	High	21	17.50

From Table 21, it could be inferred that an overwhelming majority of farm women (71.67 per cent) possessed medium level of innovativeness followed by high (17.50 per cent) and low (10.83 per cent) levels.

The results corroborated with the finding of Zamir Ahmed (1998) who concluded that most (42.50 per cent) of the farm women trainees had a high degree of innovativeness.

4.1.7. Self-confidence

'Confidence carries conviction' says a proverb. Self–confidence acquired by farm women could help in taking rational decisions in starting an enterprise. The results of the analysis of the data collected on self–confidence are incorporated in Table 22.

Table 22. Self-confidence level of the farm women (n=120)

Sr. No.	Category	Number	Per cent
1.	Low level	1	0.83
2.	Medium level	2	1.67
3.	High level	117	97.50

It is interesting to note from Table 22 that there existed high level of self–confidence with majority (97.50 per cent) of the farm women. Only a negligible percentage possessed low to moderate level of self–confidence (Fig. 3).

The medium level information source utilization and innovativeness had envisaged the farm women to possess more self–confidence.

This finding is in accordance with the findings of Sheela (2000) who reported that 80.84 per cent of the farm women possessed medium to high level of self–confidence.

4.1.8. Decision Making

Decision making is an important factor because the success of farming depends upon how well the family makes the decision. Hence, to know the extent of participation of farm women in taking decisions on farm and home activities, necessary and relevant data was collected and the results of the same are furnished in Table 23.

Table 23. Distribution of respondents based on their decision making pattern (n=120)

Sr. No.	Category	Number	Per cent
1.	Independent decision	55	45.83
2.	Joint decision with family members / husband	65	54.17

The habit of taking joint decisions in consultation with their husbands or other family members regarding the farm and home activities was seen with more than half of the respondents (54.17 per cent). The remaining 45.83 per cent of the women respondents made independent decisions.

In general, it may be interpreted that majority of the farm women were taking joint decisions by having discussion with family members. They considered such decisions as a part of family affairs due to their localized behavioural pattern. On the other hand, those who had high self-confidence might have come under the independent decision category. This finding has been supported by the findings of Ravichandran (1992), Seetha Lakshmi (1999) and Sheela (2000).

4.1.9. Level of Aspiration

To know the level of future aspirations of the farm women respondents in a familiar task, data was collected and analyzed. The results of analysis are furnished in Table 24.

Table 24. Level of aspiration of the farm women respondents (n=120)

Sr. No.	Category	Number	Per cent
1.	Low aspiration	1	0.83
2.	Medium level aspiration	4	3.34
3.	High aspiration	115	95.83

It was observed that majority of the farm women had high level of self–confidence, which showed that they possessed independency in making decisions as self confidence playing vital role in the decision making process. Those who had high aspirations necessarily had high self–confidence and this could be the possible reason for such a result.

This finding showed that they had the inclination towards any enterprise related to farming in order to enhance their farm income for better standard of living.

The result is in line with that of Arularasan (1992) who observed that a vast majority of respondents (72.50 per cent) had high level of aspiration and only 5.00 per cent belonged to low aspiration category.

4.1.10. Economic Motivation

The basic urge with every individual is to earn more, irrespective of their occupation. Such an urge prevails also among the farm women, motivating them to adopt innovative income–generating activities. The level of economic motivation that existed among the farm women respondents has been presented in Table 25.

Table 25. Categorization of respondents based on their economic motivation (n=120)

Sr. No.	Category	Number	Per cent
1.	Low	118	98.34
2.	Medium	1	0.83
3.	High	1	0.83

Table 25 denotes that almost cent per cent of women respondents (98.34 per cent) had low level of economic motivation. One respondent was seen in each of the medium and high level categories.

This finding is in conformity with the findings of Muhammal (1997) and Arunadevi (1998); however it contradicted the findings of Uma (1994) and Seetha Lakshmi (1999) who reported that majority of the respondents had high level of economic motivation.

4.1.11. Attitude Towards Self–employment

Attitude is the tendency to respond to some object, idea, situation or value. Knowing about the attitude - favourable and unfavourable, associated with respondents towards self–employment is essential for making the changes in the behaviour. Hence, this attitude was studied and the results are presented in Table 26.

Table 26. Attitude of farm women towards self–employment (n=120)

Sr. No.	Category	Number	Per cent
1.	Less favourable	35	29.16
2.	Favourable	20	16.67
3.	Highly favourable	65	54.17

The overall analysis revealed that more than half (54.17 per cent) of the respondents had highly favourable attitude towards self-employment followed by less favourable (29.16 per cent) and favourable (16.67 per cent) attitudes (Fig. 3).

This data was also influenced by the finding that the annual income of majority of the respondents was low to medium level. Also high level of aspiration coupled with more self-confidence was noticed in this report. And these factors might have been the reason for the higher favourable attitude.

The above result received support from the findings of Ravichandran (1997).

Table 27. Statement–wise analysis of respondents according to their attitude towards self–employment (n=120)

S.N.	Statements	Agree		Undecided		Disagree	
		No	%	No	%	No	%
1.	Self–employment is of much importance during the present period of extreme unemployment.	95	79.16	5	4.16	20	16.68
2.	Self–employment is an independent profession as it offers freedom.	55	45.83	26	21.67	39	32.50
3.	Self–employment helps one to become self sufficient in life.	77	64.16	10	8.33	33	27.51
4.	Self–employment is desirable since one need not expect any sanction from any official.	65	54.17	10	8.33	45	37.50
5.	Since there are ample technologies one can take–up self–employment easily.	31	25.84	24	20	65	54.16
6.	For unemployed women, self–employment is a pure profession for facing the vagaries of life.	79	65.84	10	8.33	31	25.83
7.	There is no necessity for an educated unemployed woman to go for self–employment as government jobs are meant for her.	27	22.50	19	15.83	74	61.67
8.	It is unwise to select self–employment as it needs more physical and mental efforts.	45	37.50	10	8.33	65	54.17
9.	Sound family background is a necessity for selecting self–employment.	80	66.67	13	10.83	27	22.50
10.	Agriculture is the basis for industries so selecting self–employment in agriculture is always worthy.	76	63.33	30	25	14	11.67

From Table 27, in statement 1, it could be observed that majority (79.16 per cent) of the farm women felt the need for self-employment during the present situation of extreme unemployment. And also they expressed that leisure time activities and off–farm activities help increase the income level.

One-third (32.50 per cent) of the farm women disagreed with the second statement. Even though they are independent by profession, farm women have to depend on the family members and others for their support and finance to carry out the self–employment ventures. So, they did not feel that self–employment is an independent profession.

The third statement about self–employment helping to become self–sufficient was also not agreed by 27.51 per cent of the farm women as self–employment involves more risks and uncertainties. Also their dependency on others for marketing the products might be the reason for this result. They also added that they could not expect success all the time and that failure could not even get them a minimum income.

As discussed earlier, the farm women were wholly dependent on financial institutions to get loans and on scientists / research institutes to know about scientific technologies. So, nearly half (45.00 per cent) of the respondents disagreed with the fourth statement. Over and above, the farm women expect the sanction from officials for getting a license to start an enterprise and for marketing the products.

Though numerous technologies are available, the farm women were not equipped with scientific technologies due to inadequate training and limited exposure to self–employment avenues. So, more than half (65.00 per cent) of the farm women disagreed with the fifth statement.

An entrepreneurship is not simply masculine job such as the army or navy. Women too, are equally endowed with the psychological qualities and managerial abilities that matter in successful entrepreneurship. Though self–employment requires physical efforts, apart from farming, the farm women were able to allot time to carry out the income–generating activities. It reflected on the result that more than half (54.17 per cent) of the women respondents disagreed with the eighth statement. However, more than one-third (37.50 per cent) endorsed the statement.

Choosing an enterprise should be an individual's free choice. But in most cases, it is the decision of the family and more so of the senior male member or the main earning member in the family. Hence, there is a need to stimulate and influence the whole group so that the targeted entrepreneur gets the encouragement and support to start and run the enterprise. This might be the reason why majority (66.67 per cent) of the farm women expressed that sound family background is a necessity for selecting self–employment.

And majority (63.33 per cent) of the farm women expressed their preference for agro–enterprises. Such agro–enterprises could be either input oriented (those manufacturing or processing and supplying the inputs for agriculture) or output oriented (those dealing with the output of agriculture, processing them into various end products for industrial or human consumption).

Apart from the above facts most of the woman respondents opined that time is a highly limited resource, which can neither be saved nor detained for the tasks to be performed both at home and on farm. Hence, she will have to apportion her time to meet all the responsibilities. It was observed that they worked for longer hours, in kitchen and on farm, compared to their counterparts. The energy spent by them in performing these tasks is more than it is physically feasible, particularly in a below subsistence level of living. This might be the main reason for not starting an enterprise.

4.1.12. Trainings Undergone

Scientists are generating new technologies that are useful to farmers. These technologies require more understanding and acquisition of skill on the part of the farmers to follow them. The participation of the respondents in various training programmes helps them to know the latest technologies and adopt them in their farm. The relevant data collected on trainings undergone are presented in Table 28.

Table 28. Trainings attended by farm women (n=120)

Sr. No.	Category	Number	Per cent
1.	Up to three trainings	24	20.00
2.	Four and five trainings	75	62.50
3.	More than five trainings	21	17.50

From Table 28, it could emphatically be inferred that nearby two-thirds of the farm women (62.50 per cent) attended up to five trainings followed by up to three trainings (20.00 per cent) and 17.50 per cent woman respondents attended more than five trainings.

This finding is in contrast with the findings of Uma (1994), who found that majority of the respondents did not attend any training, and Seetha Lakshmi (1999) who pointed out that majority of the respondents participated in 1–2 training programmes.

4.1.13. Perceived Effectiveness of Training

The perceived effectiveness of training enables the assessment of further needs of the respondents to adapt to a new technology or to start an enterprise. The needed data was gathered and presented in Table 29.

Table 29. Perceived effectiveness of trainings attended by the respondents (n=120)

Sr. No.	Category	Number	Per cent
1.	Less sufficient	6	5.00
2.	Sufficient	56	46.67
3.	More sufficient	58	48.33

From Table 29, it is evident that almost an equal proportion perceived the training as 'more sufficient' (48.33 per cent) and 'sufficient' (46.67 per cent). And the rest 5.00 per cent of farm women perceived the training as 'less sufficient'. It could be concluded that 95 per cent of the sample satisfied with the trainings they attended.

This finding derived support from the findings of Zamir Ahmed (1998) who reported that majority of the respondents felt 'sufficient' to 'more sufficient' towards perceived effectiveness of training.

4.1.14. Attitude Towards Training

Attitude refers to the positive or negative mental disposition of an individual towards psychological object. Multiplicity of attitudes is logically related to a person's behaviour. Attitude plays a vital role in influencing the behavioural components.

As this study primarily aimed at assessing training needs, it was felt necessary to include the variable 'attitude towards training'. The respondents were classified into three categories as less favourable, favourable and more favourable based on their attitude using cumulative frequency method. The results are given in Table 30.

Table 30 denotes that nearly three-fourths of the farm women (70.84 per cent) exhibited a highly favourable attitude towards training.

Table 30. Attitude of farm women towards trainings (n=120)

Sr. No.	Category	Number	Per cent
1.	Less favourable	19	15.83
2.	Favourable	16	13.33
3.	Highly favourable	85	70.84

It is a welcoming inference that the farm women need to have more training exposures, which would facilitate them for capacity building as an entrepreneur. Hence it may be inferred that there exists lot of scope for the trainers to organize periodical training to farm women to improve behavioural skills and technical competency.

According to Uma (1994) and Muthammal (1997) most of the farm women had favourable to more favourable attitudes towards training and thus, the result of the study is supported.

4.2. TRAINING NEED ASSESSMENT

Training is the important input in improving the professional competence of an entrepreneur viz., the farm women. Training is effective if it is based on systematic assessment of training needs.

Training is the process of changing attitudes, improving knowledge and developing skills of the persons or employees of an organization, so as to enable them to perform their jobs effectively.

Training need is the gap between what is going on now and what should go on. It is the gap between the present level of performance and the standard level of performance of job. ***- Johnson (1967)***

Training need exists when an individual lacks the knowledge or skill to perform an assigned task satisfactorily. ***- Laird (1978)***

Importance of assessing training needs

Training should be based on assessment of training needs. The participants acquire a part of training before entering into a job, but once they are set in their role and responsibilities, there occurs a gap between their overall problem–solving behaviour. This leads to the organization of the training programme.

Therefore, a training need exists in any work situation, when an actual condition of work behaviour differs from the desired condition in any aspect of organizational performance, when there is a change in the objectives of the organization or in the introduction of new programmes, practices and techniques.

Assessment of training needs is important for the employees to become more productive in their present jobs and to help them get ready for advancement in the organization.

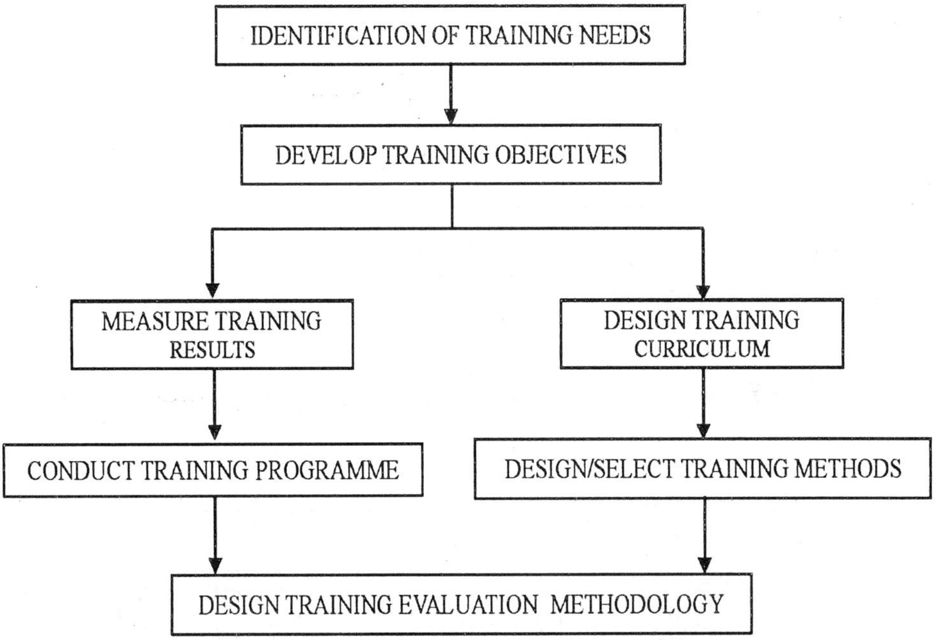

Fig. 2. Identification of training needs

The above diagram indicates the various steps involved in operating a training programme.

Assessment of training needs as a remedial measure / Training as a remedial measure

Knowledge, skill and attitude are the three criteria around which all jobs are based. A training need is a gap between the knowledge, skill and attitudes desired and already possessed by the employees. An individual needs training when his performance falls short of standards, i.e., when there is performance deficiency. Inadequacy in performance may be due to lack of skill or knowledge or any other problem, e.g., uninspiring supervision or some personal problem of the employee. The problem of performance deficiency caused by absence of skills, knowledge or attitude can be remedial for training.

Analysis performance/ Competence to suit the present /future jobs

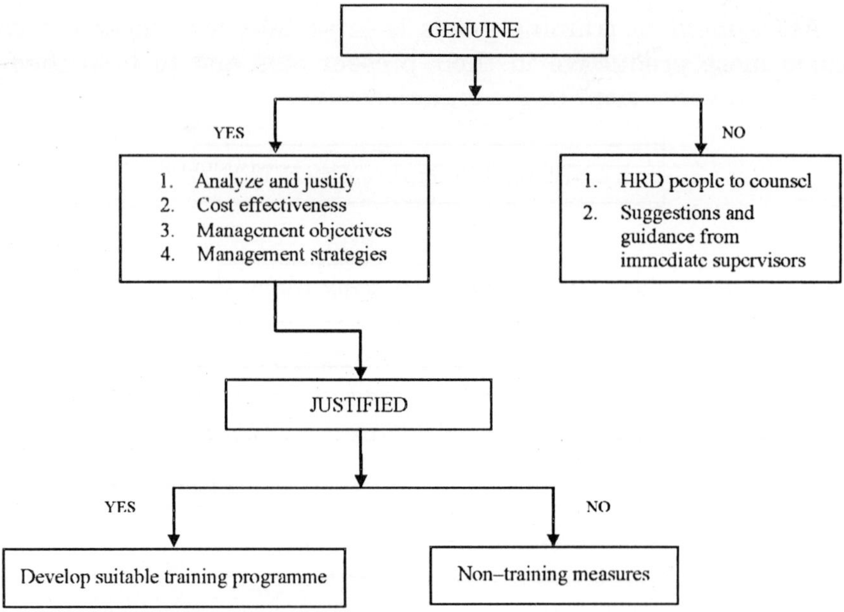

Fig. 3. Assessment of training needs and remedial measures

Types of training needs

- **Based on magnitude of content**
    - Micro training needs - These training needs pertain to specific and minute elements of curriculum content.
    - Macro training needs - A macro training need is one where a broad area of a training curriculum forms the objective of the training programme.
- **Based on number of training needs**
    - Individual training need - An individual training need exists just for one person or for a very small population.
    - Group training needs - It exists in a large group of works or frequently in the entire population with the same job classification.
- **Based on temporal utility**
    - Short range training needs - These are the training needs without whose fulfilment, one's day to day work suffers significantly.
    - Long range training needs - These are the training needs whose fulfilment increases one's professional competency

and the ability to handle the work more effectively in the coming years.

- **Based on status of individual**
  - Pre–service training needs - These training needs predominantly consist of orientation of individuals about objectives, resources, structure and functional approaches being followed by an organization.
  - In–service training needs - These training needs are primarily focused on specific knowledge, skills and attitudes desired by an individual to perform his task assigned by an organization efficiently and effectively.
- **Based on level of training needs**
  - No training needs
  - Low training needs
  - Medium training needs
  - High training needs

Methods of assessing training needs

1. Analysis of the new programme

The starting of a new programme or bringing out major modifications in the older ones, call for new knowledge, skill and attitudes on the part of the extension personnel. To prepare the extension personnel for changes, the extension organization should try to find answers to each question.

- What way the new technology or programme is different from the old one?
- What new knowledge and skill will be needed?
- When will they need it?
- Who will need it?

The answers of similar questions provide clues to the training needs. While using this method, care should be taken to get answers to the questions through various sources.

2. Analysis of an activity

This method requires analysis of level and amount of knowledge and skills needed for performing an activity by an individual. This analysis can help in assessing the training needs of an individual or a group.

3. Analysis of the problem

Training needs can be assessed by analyzing the situation being faced by the individuals of an organization. The problem may also emerge in part as an individual or a group may not have enough knowledge, skill and understanding to handle a specific challenge at a given moment in a specific situation. For analyzing the training needs, following questions are raised:

- What is the exact nature of the problem?
- Why is it so?
- What kind of attitude, knowledge and skill are missing?
- What should be done so that the problem can be handled properly?

Before starting any training program, analysis of the problem is the foremost step to carry out for any activity.

4. Self analysis method

This method includes self–analysis carried out by the people to do their best and to set high standards for themselves. People know what they need to handle their jobs through knowledge, skill and insight. These give direct clues to the training needs.

5. Knowledge test / questionnaire method

Questionnaire is developed and administered as a tool for need analysis based on structure and function of the organization and individuals, role performance, and responsibility in terms for future development of the projects. Responses in the questionnaire are evaluated and summarized to identify the training needs.

6. Interview methods

A training specialist arranges informal meeting with the concerned persons or groups to get information about the training needs. While preparing for this meeting, the training specialist develops a well–structured interview schedule conversing questions about:

- Accomplishment of job
- Improvements that could be made for better accomplishment
- Personal and career goals

- Human relations
- The need for acquiring additional knowledge, skill and attitude towards efficient services, and for handling processes, practices and products

7. Card sort method

This is a forced choice procedure. A batch of 3 x 5 cards is typed up with a single job related to single statement on each. Each statement indicates the potential training need. These cards are handed over to the persons, whose ideas are sought, to arrange these cards, in order of importance that is considered approximately by him. If an individual feels a card should not be included at all, he leaves it out. The arrangement of these cards gives clues not only about the training needs but also about the order of priority in which the training objectives need to be formulated and conducted.

8. Observation

Extension organization can employ training specialists to advise them in the matter of training of their personnel. Usually, such specialists have freedom of movement throughout the organization and they get the opportunity to observe things in the normal meetings. These observations often give clues to such training needs that are just under the surface or have started emerging. In the absence of such expert, immediate supervisor can also observe the activities of his subordinates from a relatively close distance. For doing this, he needs an alert and creative mind, good eyes, sharp ears and discreet mouth.

9. Committee method

An advisory committee comprised of responsible persons is involved in assessing the accuracy of the training needs.

10. Perception by self

A questionnaire should be developed and put before the extension personnel to fill it. They will judge their own abilities and needs as per their job level, which can help them in determining their training needs.

11. Perception by supervisors

The same questionnaire used in the 'perception by self' method should be introduced among the supervisors of extension personnel.

12. Combination of perception by self and supervisors

The results of 'perception by self and supervisors' are cross–checked and training needs of extension personnel are determined.

13. Workshop method

The workshop method involves a technique, which brings a group together to develop further skills through actual practice in a management function such as planning or report writing. As the group pursues its working goal, evidence of individual and group needs may emerge for further training in skills or knowledge areas.

14. Opinion poll

This method can be used to find out the kind of training the extension personnel need by asking them directly about it. For example, in which of the area(s) do you think, you and your subordinate need training most in order to do your job more effectively? Although many of these opinions are subject to distortion as sometimes people are unable or unwilling to respond correctly.

15. Critical incident technique

This technique can be used to assess the training needs of extension personnel. An extension functionary may be asked to recall the events during the past one or two years of his career on the extension job when he found himself particularly ineffective. After this, he may help to recall the events that took place before the incident or the cause, which according to him, made him effectively on the job. By carefully analyzing the responses in respect of such incident, one can get an idea of the training needs of the extension workers. A careful tabulation of the responses of a number of such extension personnel could help reveal the type of additional knowledge and or skill that the extension personnel need to handle the situation successfully.

16. Buzz session method

Buzzing is a group dynamic technique used to tap the thinking of a group for possible solutions to a common problem. It consists of dividing the audience into small groups of four or five persons. Each group starts to discuss the questions. The chairman of each group reports what his group has produced. Duplications are avoided and final report on training need is prepared.

17. Brain storming method

It is a useful method to assess the training needs. The procedure is simple and involves the following steps:

a. Bring together a homogenous group
b. Place in front of them a question of common interest
c. Place a problem before the group for discussion and write the ideas on a black board. One should not be judicial or try to organize the list.
d. When time is up, examine the list, and identify items which call for additional knowledge, skill or attitude.

This list provides an inventory of the training needs.

18. Performance test method

The steps of a skill are assigned some specific numerical value. The skill tester has to give the marks for a particular step and total marks obtained will indicate the training need.

A systematic training need assessment is necessary to formulate a purposeful training programme. Training need assessment should not be made only on the basis of job description but the job perceived should also be taken into consideration. The tool developed for Training Need Assessment (TNA) should cover all aspects of the needs relevant to the job requirement of the extension personnel. The training need of the extension personnel should include subject matter, communication skills, behavioural skills and managerial skills.

The methods of TNA such as knowledge test, interviews, observation, workshop, perception by self, perception by supervisor and combination of self and supervisors should be given more priority. The services of qualified and properly trained extension personnel responsible for implementing of training programmes should be utilized for assessment of training needs.

4.3. TRAINING NEEDS OF THE RESPONDENTS IN SELECTED AGRO ENTERPRISES FOR ENTREPRENEURIAL DEVELOPMENT

4.3.1. Layer Production

Poultry industry in South Asia

Poultry provides an immense supply of food for the world's population. All over the globe, poultry meat and eggs are preferred to other kinds of animal food products for a variety of reasons. It is estimated that 25.00 per cent of the world's meat supply is derived from poultry i.e., chicken, turkey, duck, geese, domesticated quail, etc. and the proportion is increasing steadily. The trend has been more noticeable in developing countries in recent years. Even though poultry meat and eggs are consumed in both developed and developing countries and this is not discouraged by the many religious taboos, the quantity of consumption has remained much lower in developing countries in comparison to developed countries. This could be partly due to eating habits as well as due to the comparatively lower purchasing power of developing countries.

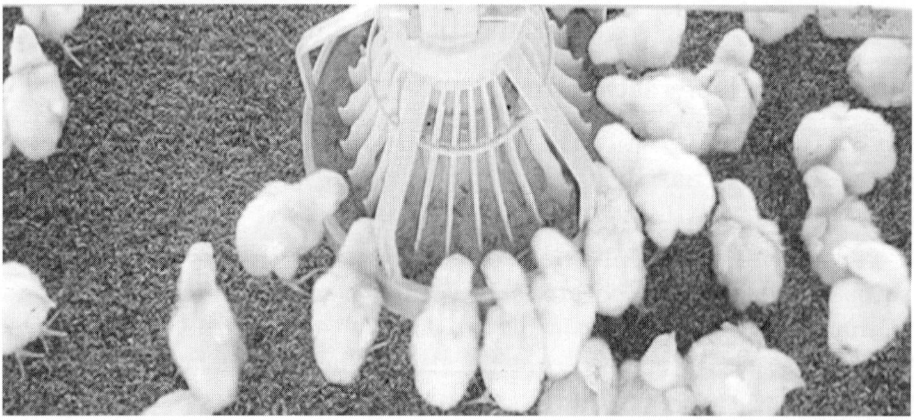

Fig 4. Layers chicks

South Asian countries are located in the tropical region of the world and the prevailing macroclimatic conditions in these countries are not the most congenial for poultry production. Yet the growing need of the ever–increasing population in the region raises the demand for poultry products. For a long time, the bulk of this demand has been met by the native breeds of chicken. The commercial poultry production with high yielding hybrid strains has only been introduced in the last 40–50 years. Most of the eggs and meat come from three types of chickens taken from such stock. The first of these are indigenous chickens that have existed in these areas for centuries, most of them living as scavengers or reared in backyards in rural conditions. Their productivity level is very low but they possess genes that are well–adapted to the tropical environment of their countries. Secondly, they have medium–level stocks, which consist mainly of pure breeds maintained by research stations and fancy breeders. These breeds are reasonably productive and are also comparatively more resistant to diseases than the hybrid chickens. Lastly, the third group of industrial stock, comprising hybrid strains, evolved out of three or four–way crosses by breeder companies in developed countries. These grandparent stocks are imported into South Asian countries by franchise breeders. Commercial strains, which rank very high in performance through random sample testing in developed countries, do not repeat the same level of superior performance in tropical South Asian countries. The franchisers in these countries could, however, identify and locate the strains, which are better–suited to local conditions, from the list of different strains of layers and broilers developed in other countries. Present–day commercial strains made available by local franchisers in these countries now have matching performance levels compared to their contemporaries in developed nations with more optimal temperate climatic conditions.

Andhra Pradesh is proving to be a frontrunner in providing farmers with direct sale of their agricultural and other products, thus enabling a maximum share of the price of the products to go to the producer farmers. Such attempts should be streamlined and farmers should be encouraged to develop the direct sale of eggs and poultry meat, especially if the native chickens are upgraded through the planned introduction of moderately superior strains as the production level of native chickens will then probably exceed local demand.

Commercial poultry production

Questions of concern are how profitable is poultry production and what conditions are needed for commercial poultry production? Poultry production (poultry meat, broiler or egg production) needs some skill, but it can be easily mastered. However, the profitability in commercial poultry production depends not only on efficient production, but also on successful marketing of the product. Anyone wishing to attempt commercial poultry production should, apart from mastering the techniques of poultry production, study the marketability of the product (poultry meat or eggs) in his region or elsewhere to ensure a reasonable profit margin. Production techniques include proper planning while selecting the location and lay–out, proper designing of poultry houses, arranging for quality inputs like chicks and feed, adopting appropriate rearing techniques and taking adequate disease–control measures to ensure high efficiency and productivity.

Profitability in poultry production also involves proper assessment of demand, planning the size of the activity, possible integration of activities to bring down the cost of production, thorough costing of production activities, foresight into market price situations, prudent assessment of cost-benefits and the rate of returns in the activity. These situations vary from country to country and even between regions of a country. The term 'Poultry' denotes all domesticated species of birds including chicken, ducks, turkeys, Japanese quail, pigeons as well as rarities such as ostrich and emu. Commercial poultry production involves various activities such as the establishment of broiler chicken farms, layer (egg production) chicken farms, rearing Japanese quail or turkeys for meat, rearing ducks for eggs, manufacturing poultry feed, establishing poultry meat shops, as well as poultry or egg trading activities etc.

How to meet the requirement?

In rural areas of Asian countries, the women have experience in rearing poultry in their backyards and in earning income from their poultry for subsistence living. The experience they possess is not sufficient to go for commercial venture. The commercialization requires much more expertise in handling of the layers from day-old chick to marketing of eggs and culled birds. They might have treated the birds for diseases with their indigenous means but, those approaches are not sufficient for the commercial production, which needs scientific means of handling them. The scientific mode of rearing of birds could be possible through skill–based training by technically–competent persons especially veterinarians. Hence, those farm women who have inclination to establish an enterprise in layer production need to be trained on the scientific means of production techniques. This will enable the women entrepreneurs to sustain successfully.

Training needs of farm women in Layer Production

Training needs can be defined in terms of gap between requirements and performance. The gap could be in terms of knowledge, skill and attitude. Training is considered to be an essential factor that has its own influence in gaining knowledge and skill in respect of the subject matter that is taught during the training.

Attempts were made to identify the specific areas in which they required training under layer production. The related findings have been presented in this section in the order of their importance.

Training needs in specific items under layer production

The training needs under various subject matter areas of layer production were assessed and presented in Table 31.

Table 31. Training needs in specific items under layer production (n=106)

S. N.	Items	Very much needed		Needed		Less needed		Not needed	
		No.	%	No.	%	No.	%	No.	%
A.	**Housing**								
1.	Cleaning and disinfecting the house	-	-	29	27.36	23	21.70	54	50.94
2.	Cleaning and disinfecting the equipments	21	19.81	23	21.70	19	17.92	43	40.57
3.	Temperature maintenance	8	7.54	23	21.70	32	30.19	43	40.57
4.	Fumigating the house	-	-	23	21.70	39	36.79	44	41.51
B.	**Feeding**								
6.	Feeding and watering	56	52.83	20	18.87	10	9.43	20	18.87
8.	Selection of breed	7	6.60	23	21.70	35	33.02	41	38.68
9.	Grading of eggs	-	-	23	21.70	44	41.50	39	36.80
10.	Storage of eggs	4	3.78	23	21.70	42	39.62	37	34.90
11.	Removal of culled birds	45	42.46	10	9.43	26	24.52	25	23.59
C.	**Litter management**								
13.	Removal of old litters, spreading of old litter, maintaining the moisture level of litter, rating of litter	-	-	8	7.54	48	45.28	50	47.18
D.	**Medical care**								
15.	Vaccination	67	63.21	35	33.01	4	3.78	-	-
16.	Providing medicated feed	73	68.87	33	31.13	-	-	-	-
17.	Disease control measures	67	63.20	39	36.80	-	-	-	-
18.	Deworming	66	62.26	20	18.87	20	18.87	-	-
19.	Delousing	61	57.54	20	18.87	25	23.59	-	-
E.	**Marketing aspects**								
21.	Selling of eggs and culled birds	45	42.46	10	9.43	26	24.52	25	23.59

The training need areas were assessed as 'very much needed', 'needed', 'less needed' and 'not needed'. It could be observed from Table 31 that more than half of the sample represented that training is 'very much needed' in the areas, namely providing medicated feed (68.87%), vaccination (63.20%), disease control measure (63.20%), deworming (62.26%), delousing (57.54%), feeding and watering (52.83%) followed by removal of culled birds, selling of eggs and culled birds (42.46% each).

Top preference of training needs was given to medical care by cent per cent of the farm women due to the fact that in poultry farming, even if one bird is infected by a certain disease, the entire flock gets affected within no time and the disease spreads like wild fire especially during rainy season; Precaution should be taken to avoid wetting of litter material which may leads to the spread of *E. coli.* Otherwise it may lead to high mortality of birds.

The next order of priority noticed was feeding as expressed by more than three-fourths (86.00 per cent) of the farm women as the cost of feed in poultry farming accounts to the extent of 70.00 per cent of the total production cost. However, preparing your own feed mixing can minimize the cost of production of feed. Therefore, adequate knowledge and skill on all dimensions of feeds and feeding are of paramount importance to the poultry growers.

The next preference was in favour of culling, since the unproductive, unhealthy weaklings in a flock consume the feed without any economic output to the farmers and bring forth economic loss.

The farmers also preferred training in the area of marketing. Since the farmers were dependent on traders for marketing, they wanted to know about the marketing strategies to realize remunerative prices. In order to get good net return, value added pocketed products may be prepared, market tie–up may be made with star hotels and super markets. This was expressed by 81.00 per cent of the farm women.

The housing contributed the indirect effect on the production and productivity of the birds. This might be the reason for the less preference of training needs in housing (27.60 per cent). This is the typical unfelt need of the farmers, which may be brought to the surface, so that the farmers can feel and realize the real need for further adoption.

It is further revealed from the study that cleaning and disinfecting the house (50.94%), maintenance of litter materials (47.18%), fumigating the house (41.51%), temperature maintenance, cleaning and disinfecting the equipment (40.57% each), selecting of breed (38.68%), grading of eggs (36.80%), and storage of eggs (34.90%) were the areas in which the farm women did not need trainings as expressed by more than one–third of them. The said areas also contribute towards the efficient management and production, and therefore, should not be ignored. Probably they would have been ignored through experience, also such areas did not involve much skill to follow, hence such response.

The outcome has been in line with the result of Peru Mathiyalagan and Subramanian (1998) who also indicated that poultry farmers required more training on medical care of the birds and feeding techniques.

4.3.2. Mushroom Production

Global scenario

Today, mushroom farming is being practiced in more than 100 countries and its production is increasing at an annual rate of 6–7%. Present world production of mushrooms is around 3.5 million tonnes as per FAO Statistics and is over 25 million tonnes (estimated) as per claims of Chinese Association of Edible Fungi.

Presently, three geographical regions - Europe, America and East Asia contribute to about 96.00 per cent of world mushroom production. With the rise in the income level, the demand for mushrooms is bound to increase in other parts of the world as well.

As per Food and Agriculture Organisation (FAO) data, the three major mushroom producing countries, viz., China, USA, and Netherland, account for more than 60.00 per cent of the world production; however share of China itself is 46.00 per cent, which is about half of the world mushroom production. China alone is reported to grow more than 20 different types of mushroom on commercial scale and mushroom cultivation has become China's sixth largest industry. China has been producing mushrooms at very low costs with the help of seasonal growing, state subsidies and capturing the potential markets in the world with processed mushrooms at costs not remunerative to the growers in other mushroom producing countries.

According to current Indian estimates, mushroom production of India is about 1 lakh metric tons, which is about 3.00 per cent of the world mushroom production. In USA and Europe major contribution towards mushroom production is by white button mushroom. In Asian countries the scenario is different and other species are also cultivated for commercial production.

BUTTON MUSHROOM

MILKY MUSHROOM

OYSTER MUSHROOM (APK 1)

PADDYSTRAW-VOLVARIELLA

BUTTON MUSHROOM
Agaricus bisporus

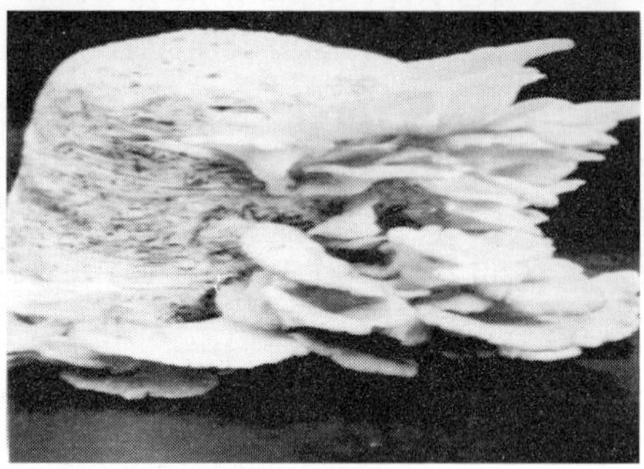

OYSTER
Pleurotus floida

MILKY MUSHROOM
Calocybe indica

OYSTER
Pleurotus.sajorcaju

Fig. 5. Types of mushrooms

Need for mushroom production

Indian agriculture will continue to be a main strength of Indian economy. With the variety of agricultural crops grown today, we have achieved food security by producing over 200 million tonnes of food grain. However, our struggle to achieve nutritional security is still on. Though we have significant achievements in milk, vegetables and fruit production, we still have to do more. In future, the ever–increasing population, depleting agricultural land, changes in environment, water shortage and need for quality food products at competitive rates are going to be important issues. To meet these challenges and to provide food and nutritional security to our people, it is important to diversify the agricultural activities in areas like horticulture.

Diversification in any farming system imparts sustainability. Mushrooms are one such component that not only impart diversification but also help in addressing the problem of quality food, health and environment related issues. One of the major areas that can contribute towards goal of conservation of natural resources as well as increased productivity is recycling of agro-wastes including agro-industrial waste. Utilising these wastes for growing mushrooms can enhance income and impart higher level of sustainability.

Major issues, which are confronting and are likely to continue to do so in foreseeable future, are: (i) population growth, (ii) income growth and (iii) urbanization. During the next 18–20 years, these factors will determine the demand for food in general, and mushrooms in particular throughout the world, especially in the developing countries. The rate of population growth is still high in developing countries as compared to developed countries. Income growth is another factor, which will not only determine the quantity but also the composition of food demand. Currently, the growth in the per capita income of developing countries is a universal phenomenon. This means that there will be increase in food consumption in most developing countries as also there will be shift in diet pattern; from staple grains to live–stock products and vegetables including mushrooms. Urbanization will also play a major role in the composition of food demand. In urban areas, the diet mostly comprises of high–value cereals, live–stock products and vegetables, a large proportion of which is in the processed form. There is a considerable migration towards urban areas in most of the developing countries and it is expected that the urban population may rise from the level of 31.00 per cent in 2005 to 57.00 per cent by 2025. In the rural areas of developing world the problems of malnutrition and poverty are alarming. Presently nearly 1.1 billion people live in poverty in the developing countries out of which about 800 million are in Asia.

It is evident from the above that the yields of major market crops of the world will have to be more than doubled, simply to maintain the current consumption pattern. In case of mushrooms, for which vast markets (like India) still remain to be tapped, the production at global level may have to be more than doubled in the next 18–20 years. Meeting the challenge of producing so much food will exert enormous pressure on the natural resource base. In spite of predominantly tropical and subtropical climates in India, it is the temperate button mushroom that has ruled and is still dominant in the mushroom scenario of the country. Taking a clue from China, it is the tropical and sub–tropical mushrooms, which should be promoted in a big way, both with the producers and consumers. A beginning has already been made with the popularization of the oyster and milky mushrooms in South India. People do develop the taste as per the available commodity, which is a universal phenomenon. At times, we take the shelter of no demand to justify almost negligible diversification in the mushroom production in India.

However, the challenges and opportunities thrown open by this unconventional crop are a bit different, particularly with reference to India. Today India is not only self–sufficient in the production of food grains, but is also in a position to export several agricultural commodities. But mushroom scenario in India is not that rosy. The cultivation of white button mushroom - *Agaricus bisporus* has come a long way of evolution and advancement in technology. Still all stages of evolution of cultivation technology can be seen. There are farmers still growing mushroom on compost prepared by the long method and there are commercial units that have shifted to bunkers or almost indoor composting technology.

Commercial production of mushroom

Even though the consumption of mushrooms has increased, there is increasing preference for fresh mushrooms. This is evident from the fact that four decades ago three–fourths of the mushrooms consumed in US were in the processed form. In 2008–09 only 15.00 per cent of the total mushrooms consumed were contributed by processed mushroom. In fact, the consumption of processed mushrooms remain more or less static in the four decades and there was about 15 times increase in the consumption of fresh mushrooms. Regular supply of energy at affordable cost is one of the constraints in commercial production of mushrooms. An important strategy to cut down the energy cost is to diversify and grow different mushrooms as per the natural conditions. No strategy for growing mushrooms can succeed if proper cultural practices are not followed. The three important Hs for mushroom cultivation are: Homogeneity, Hygiene and Humidity. For example, in button mushroom, homogeneity in wetting of substrate, Phase-I, filling of the tunnels, temperature in different corners of the tunnel, distribution of carbon-dioxide and temperature in cropping rooms is important. The lack of uniformity at any of these steps results in decrease in yield. Hygiene is perhaps the single most important issue to ensure adequate production of quality mushrooms. The disease once appeared can be managed only partially and the damage can seldom be restored. It is always proper to follow the dictum of "Prevention is better than cure". Thirdly, management of humidity (and uniformity in humidity over time and space) plays an important role to get good yields and avoid various biotic and abiotic symptoms.

Of course, along with humidity other environmental parameters like CO_2 and temperature are very important. The production becomes meaningful when there is increase in consumption / demand. To enhance local consumption there is need for popularising the cooking methods and health benefits of mushrooms.

There have been rapid changes and fluctuations in prices and demand due to global recession. The important lesson to be learnt from this scenario is that it will be appropriate to diversify in terms of different types of mushrooms, the mushroom products as well as the regions for supply of different mushrooms and products. This will provide buffering to counter the ill effects of fluctuations in prices and demand. It is possible to cultivate mushrooms under varied climatic conditions.

Commercial production of edible mushrooms bioconverts the agricultural, industrial, forestry and household wastes into nutritious food (mushrooms). Indoor cultivation of mushrooms utilizes the vertical space and is regarded as the highest protein producer per unit area and time - almost 100 times more than the conventional agriculture and animal husbandry. This hi-tech horticulture venture has a promising scope to meet the food shortages without undue pressure on land.

Post harvest practices

Mushrooms should always be harvested before water spray. The right stage for picking can be judged by the shape and size of fruit body. In young mushrooms, the edge of the cap is thick and cap margin is enrolled while the cap of mature mushroom is flat and inward curling starts. It is advisable to harvest all the mushrooms at one time from a bag so that the next crop of mushrooms starts early. After harvesting the lower portion of the stalk with adhering debris should be cut using a knife. Stipe is kept short or almost nonexistent, as it is hard and not liked by many consumers. Fresh mushrooms should be packed in perforated polythene bags for marketing. They can also be sun dried by spreading on a cotton cloth in bright sunlight or diffused light. The dried produce with 2-4% moisture can be stored for 3 to 4 months after proper sealing.

Mushroom production in India

In India the mushroom production systems are mixed type i.e., both seasonal farming as well as high–tech industry. Mushroom production in the country started in the 70s but growth rate, both in terms of productivity as well as production, has been phenomenal. In 70s and 80s button mushroom was grown as a seasonal crop in hills, but with the development of technologies for environmental controls and increased understanding of the cropping systems, mushroom production shot up from mere 5,000 tonnes in 1990 to over 1,00,000 tonnes in 2010. Today, commercially–grown species are button and oyster mushrooms, followed by other tropical mushrooms like paddy straw mushroom, milky mushroom, etc. The concentrated areas of production in India are the temperate regions for the button mushroom, tropical and sub-tropical regions for oyster, milky, paddy straw and other tropical mushrooms. Two to three crops of button mushroom are grown seasonally in temperate regions with minor adjustments of temperature in the growing rooms; while one crop of button mushroom is raised in North Western plains of India seasonally. Oyster, paddy straw and milky mushrooms are grown seasonally in the tropical or sub–tropical areas from April–October. The areas where these mushrooms are popularly grown are Orissa, Maharashtra, Tamil Nadu, Kerala, Andhra Pradesh, Karnataka and North Eastern regions of India. Some commercial units are already in operation located in different regions of our country and producing quality mushrooms for export. The present production of white button mushroom is about 85% of the total production of mushrooms in the country.

India produces about 600 million tonnes of agricultural waste per annum and a major part of it is left out to decompose naturally or burnt *in situ*. This can effectively be utilized to produce highly nutritive food such as mushrooms and spent mushroom. Substrate can be converted into organic manure or vermicompost. Mushrooms are grown seasonally as well as in state–of–art, environment–controlled cropping rooms all the year round in the commercial units. Mushroom growing is a highly labour–oriented venture and labour availability is no constraint in the country and two factors, i.e., availability of raw materials and labour make mushroom growing economically profitable in India. Moreover, scope for intense diversification by cultivation of other edible mushrooms like oyster, shiitake, milky and medicinal mushrooms are additional opportunities for Indian growers.

On the export front also, till 1993, Indian contribution to the world trade was almost negligible, but it was heartening to note that for the first time, during 1994, India not only figured in the US imports, but emerged as the second largest exporter of canned mushrooms replacing Taiwan. Now a big export– oriented unit (rated as one of the biggest in the world) and few medium–sized commercial units are exporting mushrooms to the American, European and other countries regularly. Although India is endowed with favourable natural agro–climate and a rich source of agro–wastes that could be exploited for cultivation of diverse mushroom species, yet our country does not have any significant status either as a mushroom producer or as a consumer. The per capita consumption of mushroom in India is a dismal 30–40 g as compared to 2–3 kg in USA and Europe. India itself is a big market and increase of per capita consumption even up to 100 g from present figure will help growers to market over 1.00 lakh ton mushroom within the country. People in India have become health conscious and economic conditions are good enough to permit majority of Indians to consume mushrooms as a regular vegetable on daily basis.

Presently fresh/canned mushrooms are available in all major towns/cities. By just diverting 1% of agro–wastes towards mushroom production, India can produce 3 million tonnes of mushroom and about 15 million tonnes of compost. Being an indoor crop, the commodity provides immense opportunities for empowerment of rural and urban women through cultivation and also the production of value–added products. Mushrooms possess significant health benefits and medicinal properties including anti–cancer effect. India can enter into a big and lucrative international trade in the medicinal mushrooms, presently monopolized by some East Asian countries and America. There is tremendous scope for diversifying mushroom export by including other mushroom species. With the current growth rate of the Indian economy, the domestic market too for the mushrooms is likely to enlarge sooner than later.

Oyster mushroom

Oyster mushroom, commonly referred to as 'Dhingri' in India, is a basidiomycetes and belongs to the genus '*Pleurotus*'. It is lignocellulolytic fungus that grows naturally in the temperate and tropical forests on dead, decaying wood logs, sometimes on drying trunks of deciduous or coniferous trees. It can also grow on decaying organic matter. The fruit bodies of this mushroom are distinctly shell, fan or spatula shaped with different shades of white, cream, grey, yellow, pink or light brown depending upon the species. However, the colour of the sporophores is extremely variable character influenced by the temperature, light intensity and nutrients present in the substrate. The name *Pleurotus* has its origin from the Greek word, '*Pleuro*' that means formed laterally or lateral position of the stalk or stem.

The oyster mushroom is one of the most suitable fungal organisms producing protein–rich food from various agro–wastes without composting. This mushroom is cultivated in about 25 countries of Far–East Asia, Europe and America. It is the 3rd largest cultivated mushroom in the world. China alone contributes 88.00 per cent of the total world production in oyster mushroom. The other major oyster mushroom producing countries are South Korea, Japan, Italy, Taiwan, Thailand and Philippines. At present India produces 10,000 tonnes of this mushroom annually. It is popularly grown in the States of Orissa, Karnataka, Maharashtra, Andhra Pradesh, Madhya Pradesh, West Bengal and in the North–Eastern States of Meghalaya, Tripura Manipur, Mizoram and Assam.

Advantages of growing oyster mushroom

1. Variety of substrates

Pleurotus mushroom can degrade and grow on any kind of agricultural or forest wastes, which contain lignin, cellulose & hemicellulose.

2. Choice of species

Among all the cultivated mushrooms, *Pleurotus* has maximum number of commercially–cultivated species suitable for round the year cultivation. Moreover, variation in shape, colour, texture, and aroma are also available as per the choice of consumers.

3. Simple cultivation technology

Pleurotus mycelium can grow on fresh or fermented straw and it does not require composted substrate for growth. Substrate preparation for oyster mushroom is very simple. Further this mushroom does not require controlled environmental conditions as required by *A. Bisporus* because most of the species have very wide temperature, relative humidity and CO_2 tolerance.

4. Longer shelf life

Unlike white button mushroom, the oyster mushroom fruit bodies can be easily dried and stored. Dried oyster mushrooms can be instantly used after soaking in hot water for 5–10 min or it can be used in powdered form for several preparations. Fresh mushrooms have a shelf life of 24–48 h even at room temperature.

5. Highest productivity

The productivity of oyster mushroom per unit time is very high as compared to all other cultivated mushrooms. One can harvest minimum of about 500-700 kg of fresh oyster mushroom from one ton of dry wheat or paddy straw in 45–60 d, while with the same quantity of straw only about 400–500 kg of white button mushrooms are obtained in 80–100 d (including period needed for compost preparation). Yield of this mushroom can further be increased by supplementing the substrate with suitable nitrogen source viz., soybean and cottonseed meal or by introducing high yielding cultures/strains.

The present day cultivation technology of oyster mushroom is a result of various successive steps evolved throughout the world during 20th century. A very primitive form of growing *Pleurotus* species was adopted by Lumberman in Europe during 19th century that involved collection of wood logs and stumps showing fructification in nature and keeping them in cool and moist places. First successful experimental cultivation of *Pleurotus ostreatus* was achieved in Germany by Falck in 1917. In India cultivation of *P. flabellatus* on paddy straw was reported by Bano & Srivastava in 1962 at CFTRI, Mysore.

Medicinal and nutritional value of oyster mushroom

Oyster mushrooms are 100% vegetarian and the nutritive value of oyster mushroom is as good as other edible mushrooms like white button mushroom (*A. bisporus*), shiitake (*Lentinula edodes*) or paddy straw mushroom (*Volvariella* species). They are rich in vitamin C and B complex. Protein content varies between 1.6–2.5 per cent on fresh weight basis. It has most of the mineral salts required by the human body such as potassium, sodium, phosphorus, iron and calcium. The niacin content is about ten times higher than any other vegetables. A polycyclic aromatic compound *pleurotin* has been isolated from *P. griseus,* which possesses antibiotic properties.

The foregoing discussion explicitly indicates that mushroom production is a venture, which fetches more income to the farm women in addition to their regular farm income. Particularly the oyster mushroom, which is more suitable for Indian condition, is easy to grow with low eternal inputs and limited inputs unlike the agro enterprises. Further, the mushroom has more medicinal properties and thereby its demand increases day by day. Mushroom has poor shelf life and is amenable for quick contamination and hence, the farm women who have the inclination for mushroom production must acquire training on those aspects. The farm women are already well aware of the utility value of mushroom despite the dearth of in-depth information on the production techniques. It is of immense need to train them on the various critical areas of production. So, in this study it has been attempted to unearth the areas of training need of the farm women to enable them to become successful entrepreneurs.

Marketing of mushrooms

Marketing is getting the right product, to the right people, at the right price, at the right time and in the right way. Marketing of fresh mushrooms all over the world is not very organized except the auction system in Netherlands. Producers make direct efforts to bring the produce to the super markets and the 'wholesale distributors' element is mostly missing. However, trade in the processed (canned and dried) mushrooms sector is sizeable and organized.

For effective and efficient marketing, especially export, it is necessary to understand the global trade vis-à-vis the sources of supply, potential regions of demand and consumption patterns.

Training needs in specific items under mushroom production

Table 32 discloses that all the respondents required training in the selected areas of mushroom production. However, they vary in their response from 'very much needed' to 'less needed'.

'Contamination of beds', 'sterilization and inoculation of spawn bottles', 'preventing contamination in bottles', 'preparation of substratum for mushroom bed' and 'maintaining mushroom beds in cropping room' were the areas in which cent per cent of the farm women expressed the need for training. It is worth noting that the cent per cent of the respondents felt that they needed training. This clearly shows the need for organizing a massive training programme on the felt need areas of farmers in mushroom production.

Table 32. Training needs in specific items under mushroom production

SN	Items	Very much needed		Needed		Less needed		Not needed	
		No	%	No	%	No	%	No	%
1.	Sterilization of spawn bottles	39	54.93	15	21.13	17	23.94	-	-
2.	Inoculation of spawn bottles	32	45.07	23	32.40	16	22.53	-	-
3.	Avoiding contamination in bottles	27	38.03	28	39.44	16	22.53	-	-
4.	Preparation of substrate for mushroom bed preparation	27	38.03	31	43.66	13	18.31	-	-
5.	Preparation of bed	28	39.44	30	42.25	13	18.31	-	-
6.	Avoiding contamination in beds	42	59.16	16	22.53	13	18.31	-	-
7.	Maintaining mushroom beds in cropping room	24	33.80	32	45.07	14	19.72	1	1.41
8.	Harvesting	6	8.45	26	36.62	20	28.17	19	26.76
9.	Packing	-	-	18	25.35	34	47.89	19	26.76

Mushroom cultivation involves relatively more simple techniques and small investments. Mushroom production is such activity, which can easily be undertaken especially by farm women as a subsidiary income–earning activity.

Contamination in any form or mode i.e., may be through handling, will certainly spoil the entire unit. So much care must be involved in handling every item used for mushroom production. Hence, all the selected farm women expressed the need for equipping skill in those areas.

Vijayakhader (1994) and Mamoni Das (1997) also observed from their studies that the mushroom growers required training in all the areas of production.

Harvesting and packing were the areas in which nearly three–fourths (73.24 per cent) of the farm women required training as harvesting and packing plays a very important role in preserving the produce. However 26.76 per cent of women not expressed the need for training on the above mentioned areas as those group might had possessed skill on those areas.

4.3.3. Value Addition of Fruits and Vegetables

India is the second largest producer of fruits and vegetables in the world with an annual production of around 94 million tonnes. It has the distinction of producing almost all tropical and exotic fruits and vegetables because of varied climatic conditions. Due to the short shelf life of these crops, as much as 30.00–35.00 per cent of fruits and vegetables perish during harvest, storage, grading, transport, packaging and distribution. Only two per cent of these crops are processed into value–added products. Hence, there is a need for maximum commercial utilization of fruits and vegetables and to adapt production and marketing activities as per the requirements of the world market and to cater to domestic demand, which over the past few years, has been increasing because of various socio–economic factors. If the nutritive value of the processed food products could be maintained, this sector will emerge as a major value–added food industry.

Pineapple Cleaning

Pineapple Pulp

Processing Bottling

Grapes, Squash, Jam

Fig 6. Value addition of fruits and vegetables

Fruits and vegetables are processed into a variety of products such as juices and concentrates, pulp, canned and dehydrated products, jams and jellies, pickles and chutneys etc. The extent of processing of fruits and vegetables varies from one country to another.

Historically speaking, processing of fruits and vegetables in the simplest form like pickling, sun–drying and /or making preserves has been practised in the country from very ancient times almost in every home. Since then it has been progressing fairly well and has been meeting the entire local demand and has successfully entered the exports market. Total area under fruit and vegetable cultivation is estimated at 12 million hectares, which is 7% of total cropped area in the country.

Processing units

Presently, there are approximately over 5,200 units registered under the Fruit Product Order of 1955 distributed all over the country. Most of the units fall in the cottage and/or small–scale sector. A few modern processing plants have now come up. The segment has witnessed total investments to the tune of Rs. 75 billion in the last six years.

Modern units

After the liberalization of the economic policies in the country, a few highly modern plants to produce mango pulp, tomato paste etc. in aseptic packing, have been set up and freeze drying of many fruits and vegetables including mushroom is being taken up. It is expected that in the years to come many modern, state–of–art plants shall come up.

Joint ventures

Since liberalization, proposals of industrial licenses and 100.00 per cent export–oriented units were approved and about 248 such proposals have already been implemented. The countries with which the joint ventures have been signed are USA, UK, Netherlands, Switzerland and Germany (The proposals put forward include fields such as technology transfer, financial and / or marketing tie–ups). These tie–ups include production of items like canned mushrooms, banana and mango puree, fruit concentrates, and dehydration of vegetables particularly onion. A few proposals of frozen fruit and vegetables have also been approved.

India Vs World

The extent of processing of fruits and vegetables varies from one country to another. The processing capacity as a percentage of production is as low as 2.00 per cent in India while in developed countries, this ratio is as high as 83.00 per cent.

Indian scenario of the fruit industry

India is the third largest producer of fruits in the world, (next to Brazil and USA) with an annual production of 33 million tonnes from 2.94 million ha.

Productivity (tonnes/ha)	
India	9.17
Indian share in world production	8.61
Per capita consumption (gm/d)	
India	60
Recommended	250
National requirement in million tonnes at the present level of population	53

In combination with the possibility of doubling the average yield through various development measures, the total production of fruits is expected to increase by more than four times.

In order to achieve export potential, it is suggested that a selective approach should be adopted and efforts should be concentrated on developing a few products for sale to a few markets. Based on the existing demand and potential demand, and our ability to meet them over the next few years, the following fruits have been identified as having good market potential:

- Mango
- Grapes
- Banana
- Litchi
- Exotic fruits: Chikoo, Ber, Pomegranate, Custard apple and other fruits

Presently, most of our fresh fruits are destined for West Asia with small quantities going to meet the requirements of ethnic population in UK. However, with development of refrigerated transportation systems, it could be possible to enter the more attractive markets like Singapore, Malaysia and Japan.

India's production

The production of fruits in India is estimated at 33 million tonnes or 8.00 per cent of the world production. It, therefore, ranks third after Brazil and USA, Brazil being the number one in the world. The total area under fruits is 3.3 million hectares. The important temperate fruits grown are apples, apricots, pears, peaches, cherries, etc. In the tropical and sub–tropical group, the major types of fruits are mangoes, bananas, citrus, oranges, guavas, melons, pineapples, grapes, leaches, pomegranate etc. Besides, there are many exotic fruits like starburst, avocado pears, sapota, mangosteen, olives, kiwi, which are also grown and can be commercialized.

The Indian fruit juice and pulp industry (beverage industry) has made considerable progress in the last decade, and according to trade information, it has started contributing significantly to the international trade in tropical regions. No precise data is available on domestic production of processed tropical fruits. However, it is estimated that the total annual production of fruit juices and concentrates is around 80,000 tonnes.

Exports of processed fruits and vegetables

Major product category in processed fruits and vegetables which are exported are mangoes (fresh and pulp), grapes, other fresh fruits, dried and preserved vegetables, pickles and chutneys, and fruit beverages.

Exports of fresh grapes have shown an increase of almost 28.00 per cent as compared to the last decade with UAE as a major destination. The emerging markets for fresh grapes are the Netherlands, Oman, Bangladesh and Sri Lanka.

Exports of other fresh fruits have also shown an increase by 4.00 per cent as compared to the previous years. The major destinations were Middle East (UAE, Saudi Arabia) and Bangladesh. The emerging markets for fresh fruits are Belgium, Egypt, Maldives and Italy.

Export of dried and preserved vegetables during the two decades had an increase of 37.00 per cent compared to the year 2000. The major destination for dried and preserved vegetables were: SAARC (Bangladesh, Sri Lanka, Pakistan), European Union (France, Germany, Netherlands, Poland), Asia (Japan, Malaysia), USA, Canada, Africa (Egypt, Sudan, South Africa), Middle East (UAE, Saudi Arabia, Kuwait). The emerging markets are - LAC (Argentina), Africa (Zambia, Uganda, Kenya).

Export of mango pulp has an increased from Rs. 1,965.30 million to Rs. 2,638.50 million. The major destinations were Middle East (UAE, Saudi Arabia, Kuwait). Immense potential for export of mango pulp lie in countries in European Union (Germany, UK, France and Finland), Latin America and USA.

Export from India

The main fruits that enter the export market are mangoes, grapes, apples and citrus fruits. Other fruits identified for export are bananas, sapota, litchis etc. The total exports are about 50,000 tonnes valued at Rs. 600 million. Other popular items of export are mango chutneys, pickles, fruit juices, canned and dehydrated mushrooms, frozen and canned fruits and vegetables. The main destinations are Middle East, UK, Europe and to some extent Singapore, Malaysia etc. The target is set to increase the export to 3,55,000 tonnes valued at Rs. 6,940 million.

With the commercialization of reefer containers in India the exports of mangoes, grapes, processed mushroom etc. have started to UK, Europe, Middle East, Singapore, Hong Kong etc. This mode of transport is very cost effective as there is a saving of about 50.00 per cent on freight when compared to air.

India is the 'world's key supplier' of mango puree and concentrate, with its TOTAPURI variety being a favourite blend in juices, while the premium–priced ALPHONSO is often used as a pure fruit juice or a flavouring for ice cream. Indian processors look mainly to Europe and the Middle East for sales. However, it is the industry's view that domestic consumption is also increasing.

Constraints in exports

The industry is largely of a seasonal nature, dependent on a very limited number of 2 or 3 fruits for processing, on account of which capacity utilization is very low. The industry is perennially faced with problems of availability of raw material at reasonable and uniform prices.

The vagaries of weather and the mango–bearing cycle create uncertainties in supplies and prices, which are accentuated by speculation in trade. Established exporters have articulated the need for improving the quality of our products and simultaneously building up a quality image in overseas markets, to counter the effective competition from competing sources like countries of Latin America and Mexico.

The overseas demand for Mango pulp has led to unhealthy competition among exporting interests in India, causing quoted prices to be lowered to unremunerative levels. New entrants to the trade have accentuated the problem, with established exporters.

Vegetable industry

The vegetable–growing sector comprises the production of trade in fresh and refrigerated vegetable varieties of underground (roots, tubers) and above–ground parts of herbaceous plants (fruits, stems and leaves etc.) The sector can be divided into products originating from protected vegetable–growing and from outdoor cultivation.

Global trade

The world's vegetable production is strongly concentrated in Asia, where the crops are grown for local consumption. There is scarcely any international trade in these regions. Cultivation in Europe and USA is much more commercially based and therefore there is also substantial trading at international level. More intensive industrialization and specialization in the division of work in society means an intensification of trade. The European Community's (EC) share of fresh vegetables went up from 31.00–37.00 per cent. The largest importer is West Germany with a share of 17.00 per cent followed by France and United K with seven per cent. In 1990, the import values of the none–EC countries include temporarily stabilized vegetables. The largest exporter is Netherlands with 15.00 per cent followed by Spain with 6.00 per cent. The share of the nine countries mentioned is 42.00 per cent. Here too, the dollar's fall had a great influence on the development of the export value.

Indian scenario

India is the second largest producer of vegetables in the world (after China) with an average production of 50.99 million tonnes. India also produces 18.3 million tonnes of tubers and root crops annually. It is estimated that requirements of vegetable in the country would be about two to three times more than present production. Both area and productivity have to be increased to meet the national requirement of vegetables. The important vegetables grown are potatoes, onions, green peas, cauliflower, okra, cabbage, tomatoes, eggplant, carrots, and green chillies. The estimated production is of 67.0 million tonnes grown over about 6.25 million hectare. Many exotics and luxury vegetables like gourds, pumpkins, leeks, mushrooms, asparagus and cucumbers are also grown. In a small way some organically–grown vegetables are also available. In a recent development many technically–qualified professionals have adapted to modern methods of cultivation and mechanized farming. Some glass and polyhouses to grow flowers and vegetables are also coming up.

The section discussed so far dealt with the global and Indian scenario on the production of fruits and vegetables, per capita consumption, per capita availability and potential scope for the processed fruits and vegetables. This leads to project our status as to where we stand and the destination we should reach. Further, the scope for establishing an enterprise by the farm women is also clearly visible. However, when considering the farm women the data that we have observed in this part at national and international level are beyond their limit. Keeping that scene, the actual requirements of farm women is to be identified and to be passed on to them for acceptance at their level. It could be possible that such techniques may be disseminated in the easily, and acceptable and adoptable form.

In order to achieve this attempt, hands–on training should be organized, which will ensure their participation and acquiring skill on those aspects. Gaining confidence, conviction and satisfaction are the final outcomes expected out of any hands–on experience training and it could be possible through this attempt. Table 33 projects the training needs of farm women on value addition of fruits and vegetables.

Training needs in specific items under value addition of fruits and vegetables

It is quite interesting to note from Table 33, that all the respondents expressed that there was need for training in the critical areas and none was observed under 'not needed' category. More than half of them stated that they very much needed training in the areas like adding preservatives (79.49%), sugar pulp water ratio (78.20%), testing sugar content (76.92%), artificial colour (75.65%), contamination (64.10%), adding essence (62.83%), sterilization of bottles and vessels (61.54%) and marketing outlet (55.12%). The said factors play very crucial role in the preparation of value–added products. The quantum of adding preservatives decides the keeping quality of the product. Also the sugar brix, colour etc., enrich the preference towards the value added products. In the same way preventing contamination while bottling is the most critical as that of all preserved food products, hence such an outcome.

Table 33. Training needs in specific items under value addition of fruits and vegetables (n=78)

Sr. No.	Items	Very much needed		Needed		Less needed		Not needed	
		No	%	No	%	No	%	No	%
1.	Selection of fruits and vegetables	12	15.39	48	61.54	18	23.07	-	-
2.	Washing	12	15.39	21	26.92	45	57.69	-	-
3.	Peeling	12	15.39	21	26.92	44	56.41	1	1.28
A.	**Preparation**								
1.	Pulping	14	17.95	31	39.75	33	42.30	-	-
2.	Testing sugar content (refractometer –brix)	60	76.92	16	20.51	2	2.57	-	-
3.	Preparation of sugar: pulp: water	61	78.20	14	17.95	3	3.85	-	-
4.	Preservatives	62	79.49	16	20.51	-	-	-	-
5.	Artificial colour	59	75.65	19	24.35	-	-	-	-
6.	Essence	49	62.83	29	37.17	-	-	-	-
B.	**Bottling**								
1.	Sterilization of bottles and vessels	48	61.54	30	38.46	-	-	-	-
2.	Avoiding contamination while filling	50	64.10	24	30.76	4	5.13	-	-
C	**Marketing and storage**								
1.	Marketing outlet	43	55.12	32	41.03	3	3.84	-	-
2.	Storage	30	38.46	42	53.84	6	7.70	-	-

The result has been in line with the result of Parvathi (1995) who also revealed that farm women required training in the aspects like preparation and bottling in value addition of fruits and vegetables.

Any effort taken by an entrepreneur must be linked with the market network, which enables them to sell their produce of any kind. Being farm women their exposure to market areas may be limited and they could not get such information prior to their venture. Hence, they wanted to get that information through training programmes.

The farm women in the study had been trained, prior to this investigation in various entrepreneurial avenues assessed. However they lack skill in some of the major critical areas of the enterprises. They also felt this and expressed their needs to acquire competitive skill to emerge as a successful entrepreneur. Such areas brought to limelight through this research may be considered in organizing future training programmes, which will lead the farm women towards prosperity in their enterprises.

4.4. RELATIONSHIP AND CONTRIBUTION OF INDEPENDENT VARIABLES TOWARDS TRAINING NEED IN SELECTED ENTERPRISES

4.4.1. Relationship and Contribution of Independent Variables Towards Training Needs in Layer Production (Y_1)

The relationship and contribution of factors with training needs in layer production has been furnished in Table 34.

Table 34. Simple correlation and multiple regression analysis on training needs in layer production

Sr. No.	Variable	'r' value	Partial repression co-efficient	Standard error of regre-ssion	Computed 't' value
1.	Age	0.134 NS	0.265	0.144	1.838 NS
2.	Education	0.128 NS	2.306	1.366	1.689 NS
3.	Occupation	0.012 NS	– 0.423	0.778	– 0.544 NS
4.	Annual income	– 0.421 **	– 0.446	0.092	– 4.836 **
5.	Information source utilization	– 0.566 **	– 2.340	0.577	– 4.058 **
6.	Innovativeness	0.168 NS	0.770	0.352	2.201 *
7.	Self–confidence	0.017 NS	– 0.133	0.234	– 0.570 NS
8.	Decision making	0.080 NS	0.259	2.387	0.108 NS
9.	Level of aspiration	– 0.152 NS	– 0.079	0.305	– 0.260 NS
10.	Economic motivation	– 0.024 NS	– 0.237	0.298	– 0.795 NS
11.	Attitude towards self–employment	0.014 NS	0.140	0.192	0.732 NS
12.	Trainings undergone	0.011 NS	– 1.215	1.170	– 1.038 NS
13.	Perceived effectiveness of training	0.004 NS	1.001	0.566	1.769 NS
14.	Attitude towards training	– 0.083 NS	0.005	0.758	0.007 NS

$R^2 = 0.460$; F = 6.023**

** - Significant at P = 0.01; *- Significant at P = 0.05; NS - Non–significant

Table 34 shows that the variable viz., annual income and information source utilization had significant negative correlation at 1.00 per cent level of probability. The remaining twelve variables did not show any significant relationship with training needs in layer production.

In order to find out the relative contribution of each factor towards training needs, multiple regression analysis was performed and the results are also presented in Table 34.

Table 34 further revealed that all the 14 variables had contributed to the tune of 46.00 per cent of variation, on the training needs of the farm women. The 'F' value (6.023) was found to be significant at one per cent level.

Out of the 14 variables studied, only annual income and information source utilization had negative contribution at 1.00 per cent level of significance towards the training needs of farm women whereas innovativeness exhibited positive contribution at 5.00 per cent level of significance.

The respondents were satisfied with their income from farming. So, they might not have the inclination towards layer production to get addition income. They were not interested to engage in an enterprise involving more technical skills and to invest more money in a risky enterprise like layer production. These might be the reasons for the negative contribution of annual income towards training needs in layer production.

Since the respondents were farm women, they had limited exposure with the veterinary personnel. So, they expressed the need for training in layer production. It could be the reason ascribed for the negative significant contribution of the variable, information source utilization with training needs.

4.4.2. Relationship and Contribution of Independent Variables Towards Training Needs in Mushroom Production (Y_2)

From Table 35, it could be observed that variables viz., annual income and information source utilization had positive significant correlation at 1.00 per cent level. Age and innovativeness showed negative significant association at 5.00 per cent and 1.00 per cent levels respectively. The remaining ten variables did not have any significant relationship with the training needs of farm women. So it could be observed that age, annual income, information source utilization and innovativeness had a good influence on the training needs of farm women in mushroom production.

Table 35. Simple correlation and multiple regression analysis on training needs in mushroom production

Sr. No.	Variable	'r' value	Partial repression co-efficient	Standard error of regression	Computed 't' value
1.	Age	– 0.201*	– 0.174	0.135	–1.290 NS
2.	Education	0.130 NS	2.268	1.270	1.786 NS
3.	Occupation	0.041 NS	0.547	0.729	0.751 NS
4.	Annual income	0.308 **	0.292	0.088	3.317 **
5.	Information source utilization	0.419**	1.838	0.500	3.676 **
6.	Innovativeness	– 0.340**	– 1.051	0.335	– 3.139 **
7.	Self–confidence	– 0.012 NS	0.032	0.216	0.152 NS
8.	Decision making	– 0.158 NS	– 4.435	2.259	– 1.964 NS
9.	Level of aspiration	0.138 NS	0.989	0.291	0.340 NS
10.	Economic motivation	– 0.087 NS	– 0.180	0.284	– 0.635 NS
11.	Attitude towards self–employment	– 0.081 NS	– 0.216	0.179	– 1.211 NS
12.	Trainings undergone	– 0.114 NS	– 1.487	1.113	– 1.336 NS
13.	Perceived effectiveness of training	0.170 NS	1.012	0.525	1.984 *
14.	Attitude towards training	0.001 NS	0.077	0.714	0.109 NS

$R^2 = 0.416$; F = 5.353**;

**- Significant at P = 0.01; *- Significant at P = 0.05; NS -Non–significant

In order to find out the relative contribution of each factor towards training needs, multiple regression analysis was performed and the results are also presented in Table 35.

Table 35 further revealed that all the 14 variables had contributed to the tune of 41.60 per cent of variation, on the training needs of farm women. The 'F' value (5.353) was found to be significant at one per cent level.

The factors viz., annual income, information source utilization and perceived effectiveness of training had positive significant contribution at 1.00 per cent level of significance whereas innovativeness showed a negative significant contribution at 5.00 per cent level of significance.

Increase in annual income and information source utilization would increase the training needs of farm women in mushroom production. It might be due to the high capital investment for spawn production unit and the guidance from the scientists. Similarly, increase in perceived effectiveness of training would increase the training needs of farm women. But, increase in innovativeness would decrease the training needs of the respondents in mushroom production.

4.4.3. Relationship and Contribution of Independent Variables Towards Training Needs in Value Addition of Fruits and Vegetables (Y_3)

Table 36. Simple correlation and multiple regression analysis on training needs in value addition of fruits and vegetables

S.N.	Variable	'r' value	Partial repression co-efficient	Standard error of regression	Computed 't' value
1.	Age	– 0.056 NS	0.189	0.212	0.893 NS
2.	Education	0.154 NS	4.415	1.992	2.216 *
3.	Occupation	– 0.021 NS	– 0.588	1.144	– 0.514 NS
4.	Annual income	0.163 NS	0.271	0.138	1.963 *
5.	Information source utilization	0.353 **	2.015	0.784	2.569 *
6.	Innovativeness	– 0.159 NS	– 0.970	0.526	– 1.845 NS
7.	Self-confidence	– 0.012 NS	– 0.186	0.340	– 0.548 NS
8.	Decision making	– 0.065 NS	– 0.730	3.544	– 0.206 NS
9.	Level of aspiration	0.218 *	0.693	0.457	1.517 NS
10.	Economic motivation	– 0.004 NS	– 0.185	0.446	– 0.416 NS
11.	Attitude towards self-employment	0.139 NS	0.371	0.280	1.325 NS
12.	Trainings undergone	– 0.387**	– 7.089	1.746	– 4.060 **
13.	Perceived effectiveness of training	– 0.101 NS	– 0.728	0.824	– 0.883 NS
14.	Attitude towards training	0.014 NS	1.840	1.120	1.643 NS

$R^2 = 0.3510$; F = 4.060
** - Significant at P = 0.01; * - Significant at P = 0.05; NS - Non–significant

From Table 36, it could be concluded that variables viz., information source utilization and level of aspiration had shown positively significant correlation with the dependent variable at 1.00

per cent and 5.00 per cent levels, respectively, whereas trainings undergone showed the negatively significant correlation with training needs of respondents at 1.00 per cent level in value addition of fruits and vegetables.

Table 36 further revealed that all the 14 variables contributed to the tune of 35.10 per cent of variation on the training needs of farm women in value addition of fruits and vegetables. The 'F' value (4.060) was found to be significant at 1.00 per cent level of probability.

Among the 14 variables education, annual income and information source utilization had positive contribution towards training needs of farm women at 5.00 per cent level whereas trainings undergone exhibited negative contribution at 5.00 per cent level of significance.

The result indicated that those who had higher literacy level would like to get training in this enterprise, since it involves some complicated technologies such as using refractometer (to find the brix value). Similarly, higher annual income and more exposure to information sources reflect on the training needs of farm women whereas number of trainings undergone would decrease the further training needs in future. Even though the respondents attended more trainings it might not be related with the selected enterprise.

4.5. PATH ANALYSIS

In order to find out the direct, indirect and substantial effects of independent variables on dependent variables, path analysis was performed.

Direct, Indirect and substantial effects of independent variables on training needs in layer production

It is evident from Table 37 that the factors viz., age (X_1), education (X_2), innovativeness (X_6) and decision making (X_8) had shown maximum positive direct effects on the dependent variable viz., training needs in layer production (Y_1).

The variable viz., annual income (X_4), information source utilization (X_5) and level of aspiration (X_9) had maximum indirect effects on the dependent variable viz., training needs in layer production (Y_1).

Table 37. Path analysis showing the direct, indirect and substantial effects of selected independent variables on training needs in layer production

Variable No.	Name of the variable	Direct effects	Indirect effects	Substantial indirect effects I	II	III
X_1	Age	0.11536	0.01815	– 0.02055 X_2	0.01641 X_6	0.01163 X_4
X_2	Education	0.07757	0.05056	0.07502 X_5	– 0.03056 X_1	0.01399 X_4
X_3	Occupation	– 0.04357	0.05599	0.04015 X_5	0.01154 X_4	0.01147 X_2
X_4	Annual income	– 0.32005	– 0.10087	– 0.09848 X_5	0.00504 X_6	– 0.00419 X_1
X_5	Information source utilization	– 0.47062	– 0.09528	– 0.06698 X_4	– 0.01435 X_6	–0.01237 X_2
X_6	Innovati-veness	0.09856	0.06970	0.06851 X_5	0.01921 X_1	– 0.01635 X_4
X_7	Self–confidence	0.02156	– 0.00465	0.01281 X_5	0.01165 X_1	– 0.00604 X_4
X_8	Decision making	0.04103	0.03881	0.04327 X_5	– 0.02091 X_1	0.00952 X_4
X_9	Level of aspiration	– 0.04056	– 0.11190	– 0.09418 X_5	– 0.01637 X_1	– 0.01502 X_6
X_{10}	Economic motivation	– 0.03969	0.01610	– 0.01680 X_6	– 0.00888 X_5	0.00060 X_{12}
X_{11}	Attitude towards self-employment	– 0.04094	0.05472	0.04054 X_5	– 0.01024 X_4	0.00960 X_{12}
X_{12}	Trainings undergone	– 0.06249	0.07360	0.08960 X_5	– 0.01749 X_{14}	0.00944 X_1
X_{13}	Perceived effectiveness of training	0.02277	– 0.01849	0.01548 X_5	– 0.01470 X_4	– 0.01360 X_6
X_{14}	Attitude towards training	– 0.05928	– 0.02324	– 0.02741 X_5	0.01143 X_4	0.00603 X_{11}

Among the 42 substantial indirect effects studied, 12 acted on information source utilization (X_5), 10 passed through annual income (X_4), seven routed through age (X_2) and six acted through innovativeness (X_6). Among the rest of the effects, three variables acted on education (X_2), two acted through trainings undergone (X_{12}) and one each passed through altitude towards self-employment (X_{11}) and attitude towards training (X_{14}).

It could be concluded from the results that information source utilization (X_5), annual income (X_4) and innovativeness (X_6) were identified as crucial variables on training needs in layer production.

Direct, indirect and substantial effects of independent variables on training needs in mushroom production

Table 38 revealed that the factors, namely education (X_2), annual income (X_4), information source utilization (X_5) and perceived effectiveness of training (X_{13}) had shown maximum positive direct effects on dependent variable viz., training needs in mushroom production (Y_2).

Table 38. Path analysis showing the direct, indirect and substantial effects of selected independent variables on training needs in mushroom production

Variable No.	Name of the variable	Direct effects	Indirect effects	Substantial indirect effects		
				I	II	III
X_1	Age	–0.10547	–0.09583	– 0.04178 X_6	– 0.03829 X_2	0.02814 X_8
X_2	Education	0.14454	–0.01488	– 0.04944 X_5	0.02794 X_1	– 0.01129 X_4
X_3	Occupation	0.05731	–0.01595	– 0.02646 X_5	0.02137 X_2	– 0.01228 X_{13}
X_4	Annual income	0.25838	0.04976	0.06490 X_5	– 0.01283 X_6	– 0.00675 X_{12}
X_5	Information source utilization	0.31012	0.10922	0.05407 X_4	0.03653 X_6	– 0.02304 X_2
X_6	Innovativeness	–0.25098	–0.08855	– 0.04514 X_5	– 0.02134 X_{13}	0.01320 X_4
X_7	Self–confidence	0.01189	–0.02398	– 0.01901 X_{13}	– 0.01065 X_1	0.01542 X_8
X_8	Decision making	–0.15522	–0.00317	– 0.02852 X_5	0.01912 X_1	0.01393 X_{13}
X_9	Level of aspiration	0.02679	0.11126	0.06206 X_5	0.03826 X_6	0.01497 X_1
X_{10}	Economic motivation	–0.04928	–0.03772	– 0.04278 X_6	0.01487 X_{12}	– 0.00964 X_{13}
X_{11}	Attitude towards self–employment	–0.09532	0.01469	– 0.02671 X_5	0.02142 X_8	– 0.01989 X_6

X_{12}	Trainings undergone	-0.11277	-0.00133	- 0.05910 X_5	0.01546 X_4	0.01352 X_{11}
X_{13}	Perceived effectiveness of training	0.15465	0.01509	0.03464 X_6	0.01187 X_4	- 0.01905 X_{12}
X_{14}	Attitude towards training	0.00884	-0.00763	- 0.03328 X_{12}	0.01544 X_{11}	0.01806 X_5

The variables viz., annual income (X_4), information source utilization (X_5) and level of aspiration (X_9) had maximum indirect effects on the training needs in mushroom production (Y_2).

Of the 42 substantial indirect effects, nine acted through information source utilization (X_5), seven routed through innovativeness (X_6), five each passed through annual income (X_4) and perceived effectiveness of training (X_{13}), four each acted on age (X_1) and trainings undergone (X_{12}), three routed through education (X_2), decision making (X_8) and two passed through attitude towards self–employment.

From the result, it could be concluded that information source utilization (X_5), innovativeness (X_6), annual income (X_4), and perceived effectiveness of training (X_{13}) were identified as crucial variables on trainings needs in mushroom production (Y_2).

Direct, indirect and substantial effects of independent variables on training needs in value addition of fruits and vegetables

It could be concluded from Table 39 that information source utilization (X_5), annual income (X_4), education (X_2), level of aspiration (X_9), and attitude towards training (X_{14}) had shown maximum positive direct effects on training needs in value addition of fruits and vegetables (Y_3).

The variables viz., information source utilization (X_5) and level of aspiration (X_9) had maximum indirect effects on training needs in value addition of fruits and vegetables (Y_3).

Among the 42 substantial indirect effects, 12 acted through trainings undergone (X_{12}), eight passed through information source utilization (X_5), six acted on education (X_2), four routed through innovativeness (X_6), three each passed on level of aspiration (X_9) and attitude towards training (X_{14}), two each passed through age (X_1) and attitude towards self-employment (X_{11}), and one passed through perceived effectiveness of training (X_{13}) and attitude towards self-employment (X_{11}).

Table 39. Path analysis showing the direct, indirect and substantial effects of selected independent variables on training needs in value addition of fruits and vegetables

Variable No.	Name of the variable	Direct effects	Indirect effects	Substantial indirect effects		
				I	II	III
X_1	Age	0.07702	– 0.13337	– 0.05010 X_2	– 0.02958 X_{12}	– 0.01790 X_9
X_2	Education	0.18910	– 0.03555	0.02059 X_{12}	– 0.03642 X_5	– 0.02041 X_1
X_3	Occupation	– 0.04137	0.02026	0.02796 X_{12}	– 0.01949 X_5	0.01122 X_{12}
X_4	Annual income	0.16124	0.00210	0.04781 X_5	– 0.02162 X_{12}	– 0.00827 X_2
X_5	Information source utilization	0.22846	0.12493	0.06887 X_{12}	0.02525 X_9	0.02265 X_6
X_6	Innovativeness	– 0.15557	– 0.00381	– 0.03326 X_5	0.01282 X_1	0.01002 X_{13}
X_7	Self–confidence	0.04520	0.03295	0.01898 X_{14}	– 0.01057 X_2	– 0.01035 X_{12}
X_8	Decision making	– 0.01717	– 0.04757	– 0.02101 X_5	0.01749 X_{12}	– 0.01517 X_{11}
X_9	Level of aspiration	0.12616	0.09228	0.04572 X_5	0.02860 X_{12}	0.02371 X_6
X_{10}	Economic motivation	– 0.03406	0.03052	0.04766 X_{12}	– 0.02652 X_6	0.00590 X_9
X_{11}	Attitude towards self–employment	0.10992	0.02907	0.05124 X_{12}	– 0.02273 X_{14}	0.01717 X_2
X_{12}	Trainings undergone	– 0.36137	– 0.02544	– 0.04354 X_5	0.04142 X_{14}	0.00449 X_{10}
X_{13}	Perceived effectiveness of training	– 0.07260	– 0.02883	– 0.06103 X_{12}	0.02147 X_6	0.00830 X_2
X_{14}	Attitude towards training	0.14036	– 0.12593	– 0.10663 X_{12}	– 0.01780 X_{11}	0.01330 X_5

From Table 39, it could be concluded that trainings undergone (X_{12}), information source utilization (X_5), education (X_2) were identified as crucial variables on trainings needs in value addition of fruits and vegetables.

From the above findings, it could be concluded that annual income(X_4), information source utilization (X_5) and innovativeness (X_6) were identified as crucial variables on training needs of selected enterprises.

4.6. CHALLENGES OF FARM WOMEN IN ESTABLISHING AGRO ENTERPRISES

4.6.1. Problems Experienced by Farm Women in Layer Production

The development in India would be incomplete and lopsided unless women are fully involved in the process of development. Women entrepreneurs are facing problems of shortage of raw materials, power, finance and marketing of the produce. The study was designed with the objective to know the problems faced by farm women entrepreneurs as only few studies are available on problems of women entrepreneurs.

Table 40. Issues faced by farm women in layer production (n=120)

Sr. No.	Items	Number	%
1.	High cost of feed	100	83.33
2.	Problem of disease	90	75.00
3.	High capital investment	85	70.83
4.	High price for construction of cages	80	66.66
5.	Marketing problem	80	66.66

* Multiple responses

It could be observed from Table 40 that 'high cost of feed' (83.33 per cent), 'problem of disease' (75.00 per cent), 'high capital investment' (70.83 per cent), 'high price for construction of cages' (66.66 per cent) and marketing (66.66 per cent) were perceived as major problems as expressed by more than two–thirds of the farm women.

With regard to layer production five major items were expressed as problems by more than two–thirds of the farm women.

The problem such as high cost of branded feed was expressed by majority (83.33%) of the farm women. The respondents had to depend upon the traders for branded feed. Feed is a daily input in poultry unit, which takes the lion's share in the production cost. It could be affordable to the farmers who are maintaining a large poultry unit to establish a feed mixing unit. But small and marginal farmers could not afford it due to the requirement of heavy capital investment. The next important problem was the disease incidence (75.00 per cent). Several diseases are very common in poultry birds. The most important is the Infectious Bursal Disease Virus (IBDV). The major epidemic IBDV had ravaged the poultry farms a few years ago. There were cases where people had even lost almost all the birds. Even though vaccines had been identified, due to huge growth of population of birds and poor hygiene, the virulence of the disease had also increased. Hence, the farm women (75.00 per cent) were so hesitant to carry out layer production.

The result derived the support from the finding of Venkataramaiah and Manjula (1997).

High price cages was perceived by two–thirds (66.66 per cent) of the farm women. Cage system is commonly followed for layers, which requires more investment. It is more convenient for providing water, feed, and collection of eggs. More birds can be accommodated in the cage system than the deep litter system. Also it is possible to maintain a better hygiene in cage system. But, installation of a poultry unit requires high capital investment, which includes site cost, chicks' cages, feed stock, medicines etc. Without the shed, the cage alone would cost around Rs. 50 per bird. It was felt as too costly for average farmers to afford it. Even without a cage system, starting a poultry unit itself was a capital intensive enterprise as expressed by the farm women (70.83 per cent). It required huge sheds and also materials for keeping feed, water, and collection of eggs.

Marketing of poultry products like eggs and culled birds were the problems reported by two– thirds of the sample (66.66 per cent). Those who were running poultry units used to purchase the feed from the traders on some agreement, that the cost of the feed was settled by selling the eggs to them. Thus, the farm women were getting lesser benefits, and the traders enjoying more.

Few more problems faced by farm women entrepreneurs in layer production are discussed below:

Disposal of male chicks

When chicks are hatched for egg production, only females are needed. Male layer strain chicks have no commercial value, which means that 50.00 per cent of the chicks hatched have to be killed. Their disposal raises practical and ethical issues. Methods of disposal vary from region to region and country to country. In all cases, the aim should be to ensure that every male chick is killed humanely and instantaneously.

Chick handling

Whether chicks are to be reared for meat or egg production, it is important that they are handled with care in the hatchery. After being taken from the hatching trays by hand, chicks may either be sexed and sorted manually, or placed on a conveyor from which the males (often with different sex–linked feathering rate or feather colour) are removed for slaughter. The female chicks are then placed in disposable chick boxes with perforated ventilation holes, for transport to the rearing houses.

Variation in chick or pullet sizes

The aim of the hatcheries and farms that supply pullets should be to produce birds of even body–weight and size. Variation in size can result in later problems of aggression, poor performance and injurious pecking.

Health problems faced by farm women are as given below:

Osteoporosis

Osteoporosis in laying hens is a major welfare concern. It is the progressive loss of structural bone throughout the skeleton, which results in weakened bones. Weakened bones can lead to a high number of birds suffering keel, leg and wing fractures, which are likely to be painful. Osteoporosis can also cause birds to become paralysed, which can lead to death. Loss of structural bone in hens begins around sexual maturity and continues throughout the laying period. The process is accelerated in caged systems, which prevent birds from exercising. Fleming *et al.* (2006) found significant improvements in bone strength when birds were housed in aviaries, rather than battery cages. Nutrition also appears to affect bone strength, and the effects of osteoporosis can be minimized by providing sufficient calcium, phosphorus and vitamin D in the diet. Another contributor to the severity of osteoporosis is genotype. Some genetic strains appear to be more susceptible to osteoporosis than others. It has been suggested that it is possible to select the breed free from osteoporosis while still maintaining a high egg yield, but this has not been attempted on a commercial scale.

Chronic Respiratory disease

Cyrled Teo Paralysis

Coccidiosis

New Castle Diseases

Fig. 7. Diseases

Keel fractures

One consequence of osteoporosis is that it greatly increases the susceptibility of bones to damage and fracture. In laying hens, the bone most likely to sustain a fracture is the keel bone, which can be damaged in two main ways: (i) by misjudged landings when birds are perching or nesting in a furnished environment; or (ii) when birds are handled during depopulation at the end of lay. The incidence of keel fractures caused by furnishings is higher in non–cage systems than in cage systems. In free–range and single tier aviary systems (barns), the mean prevalence of bone breakages is 65.00 per cent, 90.00 per cent of which are keel bone breaks (Wilkins *et al.* 2004). These findings are of particular importance in the European Union (EU), where conventional cages are being banned in 2012.

However, the incidence of new breaks, those caused during depopulation, is higher in conventional cage systems than in other systems. This may be attributed to weaker bones in caged birds, due to lack of exercise. Access to the birds within the cage (i.e., the size of the aperture) and the manner in which the birds are withdrawn from the cage during depopulation are critical factors in determining bone breakages.

Injurious pecking

Injurious pecking in laying hens is a major welfare concern that can spread through flocks, resulting in pain and high mortality. Injurious pecking can occur in all types of layer hen housing. In cage systems persecuted birds are unable to escape, but the problem tends to be confined to particular cages. In non–cage systems, once injurious pecking starts it can spread rapidly throughout the whole flock. Injurious pecking comprises feather pecking, vent pecking and cannibalism.

Feather pecking

Gentle feather pecking occurs when one hen pecks at the feathers of another, without pulling or removing the feathers. Severe feather pecking occurs when feathers are pulled violently or removed. The relationship between these two types of feather pecking is not clear, and they appear to have distinct risk factors. There may be a number of reasons for the onset of feather pecking, including deprivation of natural behaviours such as ground pecking (Rodenburg and Koene 2004). The inability to perform behaviours can lead to long–term frustration, which may result in arousal, aggression or fear. Any of these emotional states may increase the likelihood that a hen will start feather pecking. There are clear genetic influences on feather pecking (Rodenburg *et al.* 2008), and epidemiological studies have identified a range of important environmental risk factors. Bald patches on hens, where feathers have been removed, encourage further pecking of exposed body tissue. This has an economic impact on production, as birds lose energy and heat, and therefore consume more food. Feather pecking is likely to be very painful for the affected hens, and may lead to cannibalism. The risks of feather pecking can be reduced by feeding mash rather than pelleted diets; providing additional foraging and fibre sources such as chopped straw and vegetables; and ensuring good litter condition, to encourage birds to peck the litter rather than each other. Reducing light intensity is a short–term measure that does not address the cause of the problem.

Cannibalism

Cannibalism occurs when the flesh or blood of another individual of the same species is consumed. It is a common problem in poultry, particularly laying hens (Newberry 2004). Cannibalistic behaviour may be learned by hens, and the problem can spread rapidly throughout a flock. Cannibalism can arise as a result of severe feather or vent pecking, which often occurs due to frustration. Producers have attempted to reduce the incidence of feather pecking and cannibalism by beak trimming, which involves removal of up to two–thirds of the upper beak. This process is likely to cause pain, and does not combat the root of the problem. Some producers raise birds at low light intensities so they do not have the visibility to perform cannibalism. This has not been effective however, as the increased light levels needed to inspect birds are associated with cannibalism. Beak trimming has been banned in a number of countries, so an alternative is needed. Providing birds with enrichment such as litter for pecking may reduce frustration. It is also important to provide pullets with litter in their rearing environment. Cannibalism is also positively correlated with mineral, protein and energy deficiencies, so providing all nutritional requirements may reduce cannibalism. Selection of genetic strains that are not pre–disposed to cannibalism should also be encouraged.

Vent pecking

As with feather pecking, vent pecking can lead to cannibalism. Vent pecking is directed at the tissue around the cloaca. This may be investigative behaviour to begin with, but once established can lead to birds pecking at internal organs or tissue. The result is often death. It is, therefore, advantageous to prevent birds from viewing the cloacal areas of other birds, by ensuring that nesting areas are not brightly lit and that there are sensible perch arrangements. It is also important not to bring the flock into lay too early, as vent pecking can be triggered when small birds are encouraged to lay large eggs too early.

Emaciation

The metabolic demands of high egg production are great, and by the end of lay many hens show signs of emaciation, poor body condition and chronic stress. This can be minimized by ensuring that a good diet with adequate levels of nutrition is supplied towards the end of the laying period.

Avoiding health problems in hens

The following are some important practical tips for avoiding welfare problems:

- Avoid conventional unfurnished cages, as these cannot provide good welfare for laying hens.
- Develop plans for preventing or coping with emergencies such as equipment breakdown or fire.
- Inspect flocks at least twice a day and check individual birds, even in cage systems where it can be difficult to observe individual birds at the back of a cage. At monthly intervals, catch samples of birds to look more closely for problems such as mite infestations or vent pecking.
- Keep good records of mortality and the causes of mortality. Record spontaneous mortality separately from culling figures.
- Seek veterinary advice if birds show signs of sickness. There are many links between poor welfare and poor health/disease. Improving one can often improve the other.
- If possible, obtain birds from rearing units close to the laying farm, as this will minimize stress during transfer. The new laying flock settles more easily and early egg production is likely to be improved.
- Do not bring the flock into lay too early. Onset of lay at 17 or 18 weeks is associated with a greater risk of vent pecking than onset of lay at 19 weeks.
- Do not place perches at heights that permit one bird to peck another bird's vent. The perches at height are also associated with 'crash–landing' and subsequent bone fracture.
- The use of mash rather than pelleted feed allows the hens to spend a longer time feeding, and reduces the risk of injurious pecking.
- The provision of good, dry litter to a depth of at least 10 cm is vital for the good management of hens in non–cage systems. Keep the litter area dry and friable to reduce the risk of injurious pecking. Add fresh litter regularly and, if possible, provide hens with additional pecking materials, such as straw or other dry vegetation.
- For birds in non–cage systems, provide a raised slatted or wire mesh area separate from the litter area.
- If the birds have access to an outdoor range area, encourage them to go outside as much as possible, by providing areas of

shelter (from sun or rain) on the range. This reduces the risk of injurious pecking in the flock.

- Birds should have at least eight hours of light and at least six hours of dark in every 24 hours period, and light levels should not be less than 10 lux. In non–cage systems, consider providing brighter light over the litter areas, to encourage birds to forage and dust–bath, and lower light levels near the nest boxes and perches, to reduce the risk of vent pecking.

Extension services

The agencies doing extension services in villages like the State Department of Animal Husbandry, extension wings of State Agricultural or Veterinary Universities and non–governmental organizations, should concentrate on more effective ways to emphasize the need for vaccination against Newcastle disease, introduction of males of moderately–improved pure bred strains adapted to local conditions, routine deworming, provision of supplemental feed (protein sources, mineral sources, etc.) during breeding and laying seasons, and other appropriate technological measures that are at present lacking at the grass root level.

The rural or household type of poultry rearing increases the availability of nutritionally–superior local poultry products like egg and poultry meat, improves the income of rural households, and creates employment opportunities in rural areas. The constraints faced by the rural or backyard poultry farmer are quite different from those experienced in industrial poultry production, and measures to overcome the rural constraints need to be evolved and put into place.

With almost 60.00 per cent of the population in South Asian countries living in rural areas, with livestock and poultry rearing ensuring the sustainability of the rural poor, it is essential that rural poultry be substantially improved. Addressing the above issues will ensure that the benefits of economic growth are more widely spread, and that the poorest communities in rural areas such as agricultural landless labourers, and the small and marginal dry-land farmers, stand to gain through the employment of women and rural youths.

4.6.2. Problems Experienced by Farm Women in Mushroom Production

Marketing of mushrooms

Despite the changing currents, there is not yet much market for the processed foods and basically fresh vegetables and fruits are preferred in this country. Fresh mushrooms have very short shelf–life, cannot be transported to long distances without refrigerated transport facility and are sold in localised markets in and around production areas. The cultivation of white button mushrooms throughout the year under controlled condition is restricted to a few commercial units and 30.00–40.00 per cent of the production is being done under natural conditions during the winters. All the problems of marketing is experienced in 2–3 winter months (December–February) when more than 75.00 per cent of the annual production comes in market for sale in limited duration and market area. Farmers face the consequences of over–saturated market and are forced to sell their produce at highly unremunerative prices. Private processors, rather than coming for rescue, get tempted to avail of the situation for their benefit.

Marketing of mushrooms in India is not yet organized. It is the simple system of producers selling directly to retailer or even to consumer, which has its own limitations. Unlike the other countries where 10.00 per cent of the total cost is earmarked for marketing, we have not given marketing sufficient thought and investment. Per capita consumption of mushrooms in India is less than 50 g as against over 1 kg in various countries. There has not been any serious effort to promote the product, to strengthen and expand the market in order to increase its consumption. Mushroom is a novel food item for this country and what to ask of its flavour, texture, nutritive value, many are not aware of 'what is mushroom and whether it is a vegetarian or a non–vegetarian item?' In the coming years, there is going to be good demand for processed and fast foods.

Mushrooms may be canned to meet the demand in the off–season and in the non–producing areas. Product diversification should also be tried. Regarding the problems of sale/export of canned mushrooms, serious thought has to be given to bring down the cost of production of mushrooms and it's processing in order to compete in the international market.

Similar type of observation was made in the study area, which was expressed by the farm women respondents. The results have been presented in Table 41.

It is disclosed from Table 41 that marketing of mushroom emerged as the foremost problem as expressed by cent per cent of the respondents. Mushrooms, being highly perishable commodity with poor keeping quality, can cause great losses to growers if they are not lifted for sale immediately. Some of the suitable methods evolved commercially for increasing shelf–life of mushroom include canning, dehydration and freeze–drying. These techniques are sophisticated and costly, and are not practicable at grass root level. Therefore, it needs quick marketing. The next important problem of 'lack of financial assistance' was expressed by 83.33 per cent of the farm women. For rural women to start an enterprise, financial assistance is a pre–requisite. Though financial assistance is given by various public organizations and private agencies, they could not get it at proper time because of the long procedures in sanctioning loans.

Table 41. Areas need to be considered in mushroom production

(n = 120)

Sr. No.	Items	Number	%
1.	Marketing	120	100
2.	Financial assistance	100	83.33
3.	Non-availability of spawn	95	79.16
4.	Preservation of produce	20	16.68
5.	Pests and diseases	20	16.68

* Multiple responses

In order to encourage the entrepreneurial behaviour of farm women, they need to be organized into groups. The group approach will enable them to obtain loans through nationalized banks with less interest. The financial institutions also fulfil the guidelines and rules in giving financial support. It could be possible for the banks to issue loan to branded, organized group, which ensure prompt repayment.

The basic input in mushroom production is the spawn. It is not possible for the rural masses to produce their own spawn as it requires sophisticated and automatic machinery and other facilities. Shortage of spawn–growing laboratories is one of the main constraints to this industry. So, more than three–fourths (79.16 per cent) of the farm women expressed 'non–availability of spawn' as a constraint. 'Preservation of produce' and 'pests and disease attack' were the minor problems expressed by only 16.68 per cent of the respondents.

This finding is in line with Vijayakhader (1994) who also found that marketing the harvested mushroom was considered as a difficult task.

Special Focus

Special drives in the form of training through agencies may be given to motivate the unemployed farm youth and women, to produce spawn in their locality itself. This approach will meet the spawn requirement.

There is not much problem in the sale of fresh *Pleurotus* due to very low production but there have been problems in selling dried 'Dhingri' particularly its export where middlemen take the lion's share. Generally, the export orders are too big to be met by a single grower. *Plerutous* growers may form a cooperative society where they may pool their product and trade. Agricultural Products Export and Development Authority (APEDA) and other Central as well as State agencies would be, too willing to help them once they are assured of sufficient consignment for export for 2–3 years.

Some of the suggestions to overcome the issues involving in the marketing of mushrooms especially of white button mushrooms in India are given below:

1. Expand the market area and strengthen the demand
   a. Popularize mushrooms using Information and Communication Technology (ICT) as delicacy with nutritive and medicinal value, on mass media such as television, also through advertisements and posters.
   b. Break consumer resistance in new areas by creating awareness, by demonstrating recipes and by distributing free samples or free recipe booklet.
2. Form cooperative societies for sale
   a. Create cold storage facility
   b. Create refrigerated transport facility
   c. Create processing facility
   d. Create distributor function for big cities
3. Decrease the cost of production and bring down the sale price to boost the demand.
4. Increase the quality and aesthetic value of packaging.
5. Train retailers about handling, storage, food value and recipes.
6. Approach supermarkets, chain vegetable stores, dairy retail counters for retail sale.
7. States should fix minimum support price.
8. Public sector marketing, processing and export organizations should come forward.
9. Assured supply throughout the year at a reasonable constant price is a key to good marketing. Efforts should be made to diversify and cultivate different mushrooms throughout the

year along with cultivating some of the important mushroom during off–season under controlled condition.

10. In a limited area, say a village or a cooperative, the crops should be time–scheduled to get a reasonable, daily, uniform production to avoid glut on a day and this is required to meet the commensurate demand. One cannot ask the consumer to purchase more because all have got a peak flush during a week.

The marketing system has to be viewed as a value chain where all components are taken care of. The increased production should get translated into economic gain for mushroom producers. This is possible only when marketing is organized and demand is ensured. Considering that awareness about consumption and health benefits of numbers of mushrooms available for cultivation is limited, their demand is also less. Hence, multipronged strategy is required that means needs to expand market, increase demand, organise marketing and form cooperatives.

4.6.3. Problems Experienced by Farm Women in Value Addition of Fruits and Vegetables

Five specific issues have been put forth in the value addition of fruits and vegetables.

The results presented in Table 42 reveal that marketing was the major constraint as expressed by cent per cent of the respondents. Usually the produce of the rural women have no brand name. So, they find it difficult to sell their produce in cities and towns.

Table 42. Issues in value addition of fruits and vegetables (n=120)

Sr. No.	Items	Number	%
1.	Marketing	120	100
2.	Training	100	83.33
3.	High cost of fruits	98	81.66
4.	Financial assistance	90	75.00
5.	Seasonal availability of fruits	80	66.66

* Multiple responses

However, they sell their prepared items such as mixed fruit jam, pickles, rose essence in the local sandies and at the door step of public bar, private organizations through personal contacts.

This finding is in conformity with report of Sehrawat *et al.* (1999), who stated that marketing was considered as the major constraint encountered by the entrepreneurs.

More than three–fourths (83.33 per cent) of the farm women expressed 'lack of training' as the next major constraint in value addition of fruits and vegetables. Even though the farm women got trained, they preferred further more training in some critical areas.

The one–day on–campus training organized on 19th of every month at Home Science College & Research Institute, Madurai, may need to be extended to three different dates and possibility of conducting off–campus training in the needy villages may also need to be explored.

'High cost of fruits' and 'lack of financial assistance' were the major threats expressed by the farm women (81.66 per cent and 75.00 per cent respectively) in this enterprise. Since majority of the sample were coming under the low to medium annual income group, they could not invest more money. In one season, if fruits and vegetables were available in abundance, the rural women were not able to convert and preserve these fruits and vegetables into marketable products within a limited time. And they did not have any improved facility to store the excess produce and make it available during the period of glut. So, more than half (66.66 per cent) of the farm women expressed 'seasonal availability of fruits and vegetables' as the important constraint. Also it could be suggested that they can produce the value added products whenever the fruits and vegetables are available on cheap rates.

The need for training and guidance at all stages could be one of the most effective ways of dealing with problems of farm women that hinder taking up self–employment avenues. The major problems of credit, non–availability of inputs, and lack of marketing facilities need to be tackled through suitable means. In order to solve these issues, the farm women may be encouraged to go for such type of ventures as a group. Also they can be motivated to base tie-up with the small produces to process the vegetables and fruits at field level when the production is more. These types of approaches will enable them to run the enterprise comfortably in all aspects. They could earn some additional income from their enterprises. Such success certainly will instigate the other farm women groups or the peer group to follow the similar line of movement.

Chapter - 5

SUMMARY AND CONCLUSION

In the existing socio–economic scenario, economic independence is considered as one of the main status indicator in the society. Women, especially in rural areas, are the deprived section of the society on account of their dependence on others for economic reasons. This necessitates the generation of opportunities so as to make them economically independent. Moreover, contribution of women to national development is crucial. The process of development would be incomplete and lopsided, unless women are fully involved in it.

Many innovations have failed due to non–satisfaction of women on the perceived attributes of innovations. So, it is necessary that they should primarily be informed, trained and convinced of a technology.

Farm woman cannot emerge as a good entrepreneur, if they are not equipped with any skill, knowledge, education and information. So, research studies are essential to assess the training needs of them to become a successful entrepreneur. In order to know the training needs of farm women for entrepreneurial development, this study was taken up with the following specific objectives:

1. To assess the socio–psychological determinants of farm women
2. To ascertain the training needs of the respondents in selected agro enterprises for entrepreneurial development
3. To study the relationship between personal characteristics and the training need
4. To assess the areas need to be focussed in establishing the enterprises for their sustainability

5.1. METHODOLOGY

The study was carried out in ten villages from Thirumangalam Agricultural Division of Madurai district in Tamil Nadu with a sample size of 120 farm women selected by simple random sampling technique.

Training need was the dependent variable taken up to study the training needs of farm women for entrepreneurial development. The enterprises chosen for this study were layer production, mushroom production and value addition of fruits and vegetables. Besides this, 14 independent variables viz., age, educational status, occupational status, annual income, information source utilization, innovativeness, self–confidence, decision making, level of aspiration, economic motivation attitude towards self–employment, trainings undergone, perceived effectiveness of training and attitude towards training were selected to study their influence on the dependent variable.

The data about the socio–personal characteristics was collected with a pre–tested interview schedule. The tool for the data collection was interview schedule. The gathered data were subject to analysis using appropriate statistical tools.

The salient findings of the study have been summarized and presented here under.

5.2. SALIENT FINDINGS

5.2.1. Socio-personal Characteristics of Farm Women

Following are the major socio–personal characteristics of the farm women respondents selected for assessing training needs for entrepreneurial development on layer production, mushroom production and value addition of fruits and vegetables:

- Majority of the farm women were young aged (49.17 per cent) and an equal proportion of farm women had education up to primary and secondary (39.17 per cent in each) levels.
- Farming alone was the main occupation for 83.33 per cent of the respondents and they were equally distributed in the low (45.00 per cent) and medium income (44.16 per cent) categories.
- Medium level of information source utilization pattern (47.50 per cent) and innovativeness (71.67 per cent) was observed among the farm women respondents.
- Farm women were observed with high level of self–confidence (97.50 per cent) and aspiration (95.83 per cent) but with low level of economic motivation (98.34 per cent).
- More than half of the farm women (54.17 per cent) took joint decisions regarding farm activities.

- The respondents had highly favourable attitude towards self–employment (54.17 per cent) and towards training (70.84 per cent).
- Almost an equal proportion of farm women perceived the training they attended as more sufficient (48.33 per cent) and sufficient (46.67 per cent).

5.2.2. Training Needs in Layer Production

The respondents needed training in the subject areas such as medical care, feeding, culling and delousing.

5.2.3. Training Needs in Mushroom Production

The farm women preferred training in the subject areas such as contamination of beds, sterilization of spawn bottles, preparation of substrate for mushroom bed preparation, preparation of bed, maintaining mushroom beds in cropping room, harvesting and packing.

5.2.4. Training Needs in Value Addition of Fruits and Vegetables

The farm women expressed their need to get training in selection of fruits and vegetables, preparation, bottling, marketing and storage.

5.2.5. Relationship of Characteristics with Training Needs of Farm Women in Layer Production

Association of characters

Two factors viz., annual income and information source utilization were found to be negatively and significantly associated with the training needs of farm women.

Contributing variables

The results of regression analysis and path analysis indicated that the variables viz., annual income, information source utilization and innovativeness were the crucial and influencing variables with the training needs of farm women.

5.2.6. Relationship of Characteristics with Training Needs of Farm Women in Mushroom Production

Association of characters

Two variables, namely annual income, information source utilization had positive significant correlation whereas age and innovativeness showed negative significant association with training needs.

Contributing variables

The results of regression analysis and path analysis revealed that annual income, information source utilization, innovativeness and perceived effectiveness of training were the highly influencing variables with the training needs of farm women.

5.2.7. Relationship of Characteristics with Training Needs of Farm Women in Value Addition of Fruits and Vegetables

Association of characters

Two factors, namely information source utilization and level of aspiration had shown positively significant correlation whereas trainings undergone showed the negatively significant correlation with training needs of respondents.

Contributing variables

The results of regression analysis and path analysis indicated that the variables, namely annual income, information source utilization, trainings undergone and education were the contributing variables for training needs of farm women.

5.2.8. Constraints of Farm Women in Layer Production

The problems such as high cost of feed, disease, high capital investment, high price for construction of cages, marketing were expressed by the farm women.

5.2.9. Constraints of Farm Women in Mushroom Production

Marketing, lack of financial assistance, non-availability of spawn, preservation of produce, pests and disease were the constraints of farm women in starting mushroom production as an enterprise.

5.2.10. Constraints of Farm Women in Value Addition of Fruits and Vegetables

Marketing, lack of training, high cost of fruits, lack of financial assistance and seasonal availability of fruits were the constraints expressed by farm women in value addition of fruits and vegetables.

5.3. IMPLICATIONS OF THE STUDY

An introspective analysis of the salient findings of the study implied the following:

- The study revealed that majority of the participants preferred training on different subject areas in selected enterprises. This implied that the participants were convinced that the enterprises could generate self–employment opportunities and provide incremental income. In addition to that the farm women respondents might have understood the rationality of each of the enterprises to boost the annual income of the family, and thereby to uplift their standard of living.
- A portion of the respondents expressed that they were not at all interested to start an enterprise since it involved risks and uncertainties. Those groups can be motivated by employing suitable extension techniques like projecting success stories, training cum demonstration, symposium, seminar etc., and by involving personnel from marketing and credit institutions. The possibilities of availing credit from nationalized banks or other sector may be improved.
- Since the farm women had high level of aspiration, and high level of self–confidence, medium level information source utilization and innovativeness, they might have the inclination to start an enterprise.
- The farm women were found to be more interested in income–generating enterprises (high attitude towards self–employment). Agricultural officers may take additional effort to disseminate technologies related to such enterprises especially; veterinary personnel can be joined with them for entrepreneurial development in poultry farming for increasing the standard of living of the farm women.
- The study revealed that marketing was perceived as an important constraint in selected enterprises. Extension workers should explore the marketing avenues and create awareness among the farm women about market information on these enterprises.
- Lack of financial assistance was felt as a constraint by a portion of the farm women in starting the enterprises. This bottleneck might be overcome by organizing the farm women into self–help groups. By these self–help groups the farm women can be motivated to undertake the income–generating activities on cooperative basis.
- The study revealed that seasonal availability of fruits was also the important constraint. So, it is obvious that season–oriented trainings will be given whenever the fruits are

available at cheaper rates. Entrepreneurs may be informed on this aspect and motivated for improving their skill for further adoption.

5.4. SUGGESTIONS FOR FUTURE AREA OF RESEARCH

- A comparative study on enterprises in different locations may be conducted.
- A study on 'Rural women entrepreneurship - issues and strategies' can be taken up.
- This study focused only on three selected enterprises viz., layer production, mushroom production and value addition of fruits and vegetables. Similar study may be conducted in other enterprises.
- The study was conducted at micro level. Similar studies may be attempted at meso or macro level.

5.5 SUGGESTIONS FOR THE GROWTH OF WOMEN ENTREPRENEURS

Right efforts from all areas are required in the development of women entrepreneurs with their greater participation in the entrepreneurial activities. Entrepreneurship basically implies being in control of one's life and activities, and women entrepreneurs need to be given confidence, independence, and mobility to come out of their paradoxes. The following measures are suggested to empower the women to seize various opportunities and face challenges in business.

- There should be a continuous attempt to inspire, encourage, motivate and cooperate with women entrepreneurs.
- An awareness programme should be conducted on a mass scale with the intention of creating awareness among women about the various areas to conduct business.
- Attempts should be made to enhance the standard of education of women in general as well as making effective provisions for their training, practical experience and personality development programmes, to improvise their overall personality standards.
- Organize training programmes to develop professional competencies in managerial, leadership, marketing, financial, production process, profit planning, maintaining books of accounts and other skills. This will encourage women to undertake business.

- Vocational training should be extended to women community to enable them to understand the production process and production management.
- Skill development should be done in women polytechnics and industrial training institutes. Skills are put to work in training–cum–production workshops.
- Educational institutes should tie up with various government and non–government agencies to assist in entrepreneurship development mainly to plan business projects.
- International, national, local trade fairs, industrial exhibitions, seminars and conferences should be organized to help women to facilitate interaction with other women entrepreneurs.
- Women in business should be offered soft loans and subsidies for encouraging them into industrial activities. The financial institutions should provide more working capital assistance both for small–scale and large–scale ventures. The provision of microcredit system and enterprise credit system should be made available to the women entrepreneurs at local level.
- The weaker section could raise funds through various schemes and incentives provided by the government to develop entrepreneurs in the State, for example, the Prime Minister's Rozgar Yojana, The Khadi and Rural Village Industries Scheme etc.
- In the initial stages, women entrepreneurs may face problems but they must persevere, believe in themselves and not give up mid-way.
- Attempts should be made by various NGO's and government organizations to spread information about policies, plans and strategies on the development of women in the field of industry, trade and commerce. Women entrepreneurs should utilize the various schemes provided by the Government.
- Women should try to upgrade themselves in the changing times by adapting to the latest technology benefits. Women must be educated and trained to acquire the skills and knowledge in all the functional areas of business management. This can facilitate women to excel in decision making process and develop a good business network.

- Self–help groups of women entrepreneurs can mobilize resources and help in pooling capital funds, in order to help the women in the field of industry.
- Women entrepreneurship must be examined both at the individual level (i.e. the choice of becoming self–employed) and at the firm level (the performance of women–owned and managed firms) in order to fully understand the differences between men and women entrepreneurship.
- To establish all India forums to discuss the problems, grievances, issues, and filing complaints against constraints or shortcomings towards the economic progress path of women entrepreneurs and giving suitable decisions in the favour of women entrepreneurs and taking strict stand against the policies or strategies that obstruct the path of economic development of such group of women entrepreneurs.

Thus, by adopting the aforesaid measures in letter and spirit, the problems associated with women can be solved. Entrepreneurship is not a bed of roses to anyone, especially to women. Women participation in many kinds of economic activities to complement to their family income, no way reduces their family duties. The task of women has become more tedious and full of challenges. Let us all make efforts to help women rediscover her.

5.6. INITIATIVES TAKEN BY THE GOVERNMENT

Development of women has been a policy objective of the government since independence. Before the 70s the concept of women development was mainly welfare oriented, but during 70s, there was a shift from welfare approach to development approach that recognized the mutually–reinforcing nature of the process of development. The 80s adopted a multidisciplinary approach with an emphasis on three core areas of health, education and employment. Women were given priorities in all the sectors including Small Scale Industry (SSI) sector. Government and non–government bodies have paid increasing attention to economic contribution of women through self–employment and industrial ventures. The First Five–Year Plan (1951–56) envisaged a number of welfare measures for women. Establishment of the Central Social Welfare Board, Organization of Mahila Mandals and the Community Development Programmes were a few steps in this direction. In the Second Five–Year Plan (1956–61), the empowerment of women was closely linked with the overall approach of intensive agricultural development programmes. The Third and Fourth Five–Year Plans (1961–66 and 1969–74) supported female education as a major welfare measure. The Fifth Five–Year Plan (1974–79) emphasized training of women, who were in need of income and protection. This plan coincided with International Women's Decade and the submission of 'Report of the Committee on the Status of Women in India'. In 1976, Women's Welfare and Development Bureau was set up under the Ministry of Social Welfare. The Sixth Five–Year Plan (1980–85) saw a definite shift from welfare to development. It recognized lack of access for women to resources as a critical factor impending their growth. The Seventh Five–Year Plan (1985–90) emphasized the need for gender equality and empowerment. For the first time, emphasis was placed upon qualitative aspects such as inculcation of confidence, generation of awareness with regards to rights and training in skills for better employment. The Eight Five-Year Plan (1992–97) focused on empowering women, especially at the grass root level through Panchayati Raj Institutions. The Ninth Five–Year Plan (1997–2002) adopted a strategy of Women's Component Plan, under which not less than 30.00 per cent of funds and benefits were earmarked for women related sectors. The Tenth Five–Year Plan (2002–07) aims at empowering women through translating the recently adopted National Policy for Empowerment of Women (2001) into action and ensuring survival, protection and development of women and children through rights–based approach. At present, the Government of India has over 27 schemes for women operated by different departments and ministries.

Some of these are:

- Integrated Rural Development Programme (IRDP)
- Khadi And Village Industries Commission (KVIC)
- Training of Rural Youth for Self-Employment (TRYSEM)
- Prime Minister's Rojgar Yojana (PMRY)
- Entrepreneurial Development Programme (EDPs)
- Management Development Progammes
- Women's Development Corporations (WDCs)
- Marketing of Non–Farm Products of Rural Women (MAHIMA)
- Assistance to Rural Women in Non–farm Development (ARWIND) schemes
- Trade Related Entrepreneurship Assistance and Development (TREAD)

- Working Women's Forum
- Indira Mahila Yojana
- Indira Mahila Kendra
- Mahila Samiti Yojana
- Mahila Vikas Nidhi
- Micro Credit Scheme
- Rashtriya Mahila Kosh
- SIDBI's Mahila Udyam Nidhi
- Mahila Vikas Nidhi
- SBI's Sree Shakti Scheme
- NGO's Credit Schemes
- Micro and Small Enterprises Cluster Development Programmes (MSE-CDP)
- National Banks for Agriculture and Rural Development's Schemes
- Rajiv Gandhi Mahila Vikas Pariyojana (RGMVP)
- Priyadarshini Project - A programme for Rural Women Empowerment and Livelihood in Mid Gangetic Plains
- Exhibitions for women, under promotional package for Micro and Small enterprises approved by CCEA under marketing support

The efforts of government and its different agencies are ably supplemented by NGOs that are playing an equally important role in facilitating women empowerment. Despite concerted efforts of governments and NGOs there are certain gaps. Of course, we have come a long way in empowering women yet the future journey is difficult and demanding.

5.7. CONCLUSION

It can be said that today we are in a better position wherein women participation in the field of entrepreneurship is increasing at a considerable rate. Efforts are being taken at the economy level with promises of equality of opportunity in all spheres to the Indian women. Equal rights of participation in political process and equal opportunities and rights in education and employment are enacted. But unfortunately, the government–sponsored development activities have benefited only a small section of women i.e., the urban middle–

class women. Women sector occupies nearly 45.00 per cent of the Indian population. At this juncture, effective steps are needed to provide entrepreneurial awareness, orientation and skill development programs to women. The role of women entrepreneurs in economic development is also being recognized and steps are being taken to promote women entrepreneurship. Resurgence of entrepreneurship is the need of the hour emphasizing on educating women strata of population, spreading awareness and consciousness amongst women to outshine in the enterprise field, making them realize their strengths, and important position in the society and the great contribution they can make for their industry as well as the entire economy. Women entrepreneurship must be moulded properly with entrepreneurial traits and skills to meet the changes in trends, challenges global markets and also be competent enough to sustain and strive for excellence in the entrepreneurial arena. If every citizen works with such an attitude towards respecting the important position occupied by women in society and understanding their vital role in the modern business field too, then very soon we can pre–estimate our chances of out–beating our own conservative and rigid thought process, which is the biggest barrier in the development process of our country. We have always viewed that a smart woman can pick up a job any day, but if she becomes an entrepreneur she can provide a livelihood to 10 more women at least..!! Highly educated, technically sound and professionally qualified women should be encouraged for managing their own business, rather than being dependent on wage employment outlets. The unexplored talents of young women can be identified, trained and used for various types of industries to increase the productivity in the industrial sector.

REFERENCES

Abdul Kalam APJ 2003 Empowerments will help tackle terrorism, *The Hindu* dt: Sep. 4

Agarwal VK 1975 *Initiative Enterprise and Economic Choice in India*, New Delhi: Manoharlal Pub. Pvt. Ltd.

Alagesan V 1990 Training needs of mango growers, *Tamil Nadu Journal of Extension Education* **1** 105–111

*Alagesan V, Padmanabhan NR, Renganathan G and Henry Louis I 1990 Research Project Report on “An assessment of training needs of grape growers”, Regional Research Station, Tamil Nadu Agricultural University, Dharmapuri Dt.

Amaresh V 2004 Where women stand today, *The Hindu* dt: Mar. 25

Amudha S and Veerabhadraiah V 2000 Women in commercial poultry, *Kurukshetra* **5** 37–40

Anand J 1998 Micro Finance in Kerala, *Kurukshetra* **48** 16–21

Anandavalli M 2004 Social empowerment, *The Hindu* dt: Oct. 25

Anjaneyaswamy G 1992 Implications of Rural Entrepreneurship in Non-Farm Activities, *Kurukshetra* **11** 24–26

Arularasan GS 1992 ex–post facto Analysis of Self-Employment Training on Vocational Settlement Pattern of Rural Youths, M.Sc. (Ag.) Thesis, AC&RI, TNAU, Madurai

Arunadevi M 1998 Critical analysis on factors contributing for participation of TANWA trainees in farming, M.Sc. (Ag.) Thesis, AC&RI, TNAU, Madurai

Ayyadurai G 1980 Spatial distribution of poultry farms and facilities offered by Tamil Nadu Poultry Development Corporation, M.Sc. (Ag.) Thesis, TNAU, Coimbatore

Babykumari P 1998 Farmers and farm women in farm and home decision making, Ph.D. Thesis, TNAU, Coimbatore

Balu V 2001 Entrepreneurial development of India - Analysis of some key factors, *Social Welfare* **55**: 45-51

Baporikar N 2007 *Entrepreneurship Development & Project Management*, Mumbai: Himalaya publishing House Pvt. Ltd.

Beegam RS and Sarngadharan M 1994 Female Entrepreneurship in Kerala, *Yojana* **38** 29–30

*Blalock HM 1960 *Social statistics*, New York: Mc Graw Hill Book Company

Brush C and Taori K 1997 Entrepreneurship in the centralised Sector Women–Owned Businesses: Obstacles and Opportunities, *Journal of Developmental Entrepreneurship* **15** 157–161

Brush C and Welter F 2007 Advancing a Framework for Coherent Research on Women's Entrepreneurship, *Entrepreneurship Theory and Practice* **31** 323–39

Cantzler I and Leijon S 2007 Team-Oriented Women Entrepreneurs: A Way to Modern Management, *Journal of Small Business and Enterprise Development* **14** 732–746

Chandrakandan K, Natrajan K, and Veeramani A 1994 Impact of oil seeds campaign and training, *Journal of Extension Education* **5** 849–853

Chatterjee A 1992 Entrepreneurship Development Programme and Self-employment, *Yojana* **38** 12–15

Chidambaram K and Thenmozhi G 1998 Constraints for women entrepreneurs, *Social Welfare* **45** 28–31

Chockalingam N 1994 Role of women in dry farming system, M.Sc. (Ag.) Thesis, AC&RI, TNAU, Madurai

Desai V 1995 *Small Scale Industries and Entrepreneurship,* Mumbai: Himalaya Publishing House Pvt. Ltd.

Desai V 1999 Dynamics of Entrepreneurial Development and Management (3rd edition), Mumbai: Himalaya Publishing House Pvt. Ltd.

Dhameja SK 2002 *Women Entrepreneurs: Opportunities, Performance and Problems*, New Delhi: Deep and Deep Publishers Pvt. Ltd.

Desmarai AA 2002 The Via Campesina: Consolidating an international peasant and farm movement, *Journal of peasant studies*, **44** 554

DeTienne D and Chandler G 2007 The Role of Gender in Opportunity Identification, *Entrepreneurship Theory & Practice* **31** 365–386

Dzisi S 2008 Entrepreneurial Activities of Indigenous African Women: A Case of Ghana, *Journal of Enterprising Communities: People and Places in the Global Economy* **2** 254–264

Desigavinayagam P 1994 A study on on-campus training programme of Krishi Vigyan Kendra, M.Sc. (Ag.) Thesis, AC&RI, TNAU, Madurai

Devasena C 1992 Rural Telecast - An information Source for rural women in dry farming area, M.Sc. (Ag.) Thesis, TNAU, Coimbatore

Dharmalingam P 1990 Weedicide technology diffusion among rice farmers, M.Sc. (Ag.) Thesis, AC&RI, TNAU, Madurai

Dixit BK 1988 Small Scale Industries - Ills and Remedies, *Yojana* **32** 1–15

DSIR (Department of Scientific and Industrial Research) 2008 Report on fruits and vegetables sectors – An overview (retrieved from www.dsir/gov.in/reports), New Delhi

Eddleston K and Powell G 2008 The Role of Gender Identity in Explaining Sex Differences in Business Owners Career Satisfier Preferences, *Journal of Business Venturing* **23** 244–256

*Farr-Wharton R and Brunetto Y 2007 Women Entrepreneurs, Opportunity Recognition and Government-Sponsored Business Networks: A Social Capital Perspective, *Women in Management Review* **22** 187–207

Rashid A 2011 Female Entrepreneurs, 142–149. *In* U.A. Muhammad [ed.], *The Proceedings of the 2nd International Conference on Business Management*, Karachi

Farr-Wharton R and Brunetto Y 2009 Female Entrepreneurs as Managers - The Role of Social Capital in Facilitating a Learning Culture, *Gender in Management: An International Journal* **24** 14–31

Fleming RH, McCormack HA, McTeir L, and Whitehead CC 2006 Relationships between genetic, environmental and nutritional factors influencing osteoporosis in laying hens, *British Poultry Science* **47** 742–755

Gangadharan KK 1996 Adoption of improved agricultural practices by Pepper growers of Idukki District, M.Sc. (Ag.) Thesis, College of Agriculture, Vellayani

Gartner WB 1990 What are we Talking about When we Talk about Entrepreneurship?, *Journal of Business Venturing* **5** 15–28

Gartner WB, Shaver KG, Carter NM, and Reynolds PD 2004 *The Handbook of Entrepreneurial Dynamics*, Thousand Oaks: Sage Publications

Gonslaves 2004 Empowerment of Women, *The Hindu* dt: Feb. 20

Gordon E Schemes and Programmes of Ministry of Small Scale Industries and Ministry of Agro & Rural Industries, Govt. of India, Kalyani Publishers

Govindappa GT 1999 Rural Women Entrepreneurship-Constraints and strategies, *Kurukshetra* **4** 11–14

Gupta MP 1982 Training needs of farmers in Himachal Pradesh, *Indian Journal of Extension Education* **18** 69–71

Helen S, Perumal G, and Netaji Seetharaman R 1992 Utilization of information sources by dry farming small farm families, *Tamil Nadu Journal of Extension Education* **5** 413

Himachalam D 1990 Entrepreneurship Development in Small Scale sector, *Yojana* **34** 16–18

Halkias D, Nwajiuba C, Harkiolakis N, and Caracatsanis SM 2011 Challenges Facing Women Entrepreneurs in Nigeria, *Management Research Review* **34** 221–235

Hampton A, McGowan P, and Cooper S 2011 Developing Quality in Female High Technology Entrepreneurs' Networks, *International Journal of Entrepreneurial Behaviour & Research* **17** 588–606

Hemantha K 2001 The role of university in empowering women, *University News* **39** 9–12

Itani H, Sidani YM, and Baalbaki I 2011 United Arab Emirates Female Entrepreneurs: Motivations and Frustrations, Equality Diversity and Inclusion, *An International Journal* **30** 409–424

Jamali D 2009 Constraints and Opportunities Facing Women Entrepreneurs in Developing Countries: A Relational Perspective, *Gender in Management: An International Journal* **24** 232–251

Jeyalakshmi 1996 Entrepreneurial behaviour of Rural women in Thiruvananthapuram District, M.Sc. (Ag.) Thesis, College of Agriculture, Vellayani, Thriruvananthapuram

Jeyalakshmi G, Shilaja S, and Sobhana G 1997 Constraints experienced by women Entrepreneurs in Kerala, *Journal of Extension Education* **8** 1752–1754

Jeyasree R 1993 Impact of TANWA training on farm women, M.Sc. (Ag.) Thesis, AC&RI, TNAU, Madurai

Johnson 1967 Determining Training Needs, in *Training and Development Hand-book* (eds) RL Craig and LR Bittel, New York: McGraw-Hill Publications, 123–138 pages

Kabeer N 2001 Reflections on the Measurement of Women's Empowerment, in *Discussing Women Empowerment: Theory and Practice* (ed.) Sisack A, Stockholm: Sida, 17-27 pages

Kalivaradhan J 1990 Participation of women in IRDP - An analytical study, M.Sc. (Ag.) Thesis, TNAU, Coimbatore

Karthikeyan C 1997 An experimental study to develop an effective training model for potential growers of export-oriented cut flowers, M.Sc. (Ag.) Thesis, TNAU, Coimbatore

Kerlinger FN 1973 *Foundations of Behavioural Research*, New York: Hoth, Rinehart and Winston, Inc.

Krishna A 2003 Working Paper on Social capital, community driven development, and empowerment: A short note on concepts and operations, World Bank, Washington

Kumar A 2004 Financing Pattern of Enterprises Owned by Women Entrepreneurs, *The Indian Journal of Commerce* **57**

Kumar A 2006 Self-help groups, women's health and empowerment: Global thinking and contextual issues, *Jharkhand Journal of Development and Management Studies* **4** 2061–2079

Kothari CR 1985 *Research methodology: Methods and techniques*, New Delhi: Wiley Eastern Ltd.

Lakshmi Raju D, Nataraju MS, and Murthy N 1994 Women in animal production - An ex-post facto analysis, *Agriculture Extension Review* **5** 3–8

* Langowitz N and Minniti M 2007 The Entrepreneurial Propensity of Women, *Entrepreneurship Theory & Practice* **31** 341–364

Mahanty SK 2005 *Fundamentals of Entrepreneurship*, India: Prentice Hall of India

Maheswari M 2000 Training needs of farmers in organic farming, M.Sc. (Ag.) Thesis, AC&RI, TNAU, Madurai

Malhotra A, Schuler SR, and Boender C 2002 Measuring women's empowerment as a variable in international development, *Background paper for World Bank workshop on poverty and gender: New Perspectives, World Bank*

Das M 1997 Mushroom cultivation for rural women *Social Welfare* **44** 34–35

Mansingh JP 1993 Construction and standardization of socio-economic status scale, Ph.D. Thesis, TNAU, Coimbatore

*Maricle and Birkenholz 1988 Rural Entrepreneurship Research, *The Agricultural Education Magazine* **61** 25–28

Mary ML, Annamalai R, and Muthiah M, 1993 Evaluation of farm youths (girls) respectively to various information sources, *Journal of Extension Education* **4** 636–638

*Mathai S 1978 Paper on Risks of Entrepreneurs in Developing New Entrepreneurship, Entrepreneurial Development Institute of India, Ahmedabad

Mayoux L 2006 Participatory learning for women's empowerment in microfinance programmes: Negotiating complexity, conflict and change, *IDS Bulletin* **29** 39–50

Meena A and Puspa G 2005 Gender Indicators of equality, inclusion and poverty reduction, *Economic and political weekly* dt: Oct. 29

Goyal M And Jai Prakash 2011 Women Entrepreneurship In India-Problems And Prospects, *ZENITH International Journal of Multidisciplinary Research,* **1** 195

Microwatch A Quarterly Update on Microfinance Outreach in Pakistan"; in *Proceedings of 2nd International Conference on Business Management,* March 28-29, 2012, Lahore.

Miller NJ, Besser TL, and Riibe JV 2007 Do Strategic Business Networks Benefit Male-and Female-Owned Small Community Businesses, *Journal of Small Business Strategy* **17** 53–74

Moore D and Buttner H 1997 *Female Entrepreneurs: Moving Beyond the Glass Ceiling*, Thousand Oaks, CA: Sage Publications

Moore DP, Moore JL, and Moore JW 2011 How Women Entrepreneurs Lead and Why they Manage that Way, *Gender in Management: An International Journal* **26** 220–233

Mordi C, Simpson R, Singh S, and Okafor C 2010 The Role of Cultural Values in Understanding the Challenges faced by Female Entrepreneurs in Nigeria, *Gender in Management: An International Journal* **25** 5–21

Mustafa Z and Ismailov N 2008 Entrepreneurship and Microfinance: A Tool for Empowerment of Poor - Case of Akhuwat-Pakistan, Master's Thesis,

School of Sustainable Development of Society and Technology, Malardalen University

Mohanthy M 1994 Decision making in home and farm related activities, *Kurukshetra* **13** 37–40

Mohiuddin A 1987 Entrepreneurship in Rural Women - A study, *Kurukshetra* **35** 19–25

Banerjee M and Talukdar RK 1997 Variables influencing entrepreneurship of women entrepreneurs, *Indian Journal of Extension Education* **33** 25–33

Muhammad AS and Al-Ghazali BM 2012 The Dynamics Of Entrepreneurial Motivation Among Women: A Comparative Study Of Businesswomen In Saudi Arabia And Bahrain, *Asian Academy of Management Journal,* **17** 97–113

Muhammad UA and Gul Rashid A Female Entrepreneurs, *Proceedings of 2nd International Conference on Business Management* March 28-29, 2012, Lahore.

Murugesan P 1996 Impact of training on mushroom cultivation-An analysis, M.Sc. (Ag.) Thesis, TNAU, Coimbatore

Muthammal K 1997 Training needs of farm women in rice cultivation, M.Sc. (Ag.) Thesis, AC&RI, TNAU, Madurai

Nabi K and Mohanty AK 1992 Entrepreneurship Development in Orissa-Some Issues, *Yojana* **36** 18–21

Nadgrodkiewicz A 2011 Empowering Women Entrepreneurs: The Impact of the 2006 Trade Organizations Ordinance in Pakistan, *Center for International Private En*terprise

*Nadkarni S 1988 Abstract on Women Entrepreneur's socio-economic study with reference to Poona

Narmatha 1990 Involvement of women in poultry farming activities, *Journal of Extension Education* **6** 1188–1190

Natarajan K 2007 *Entrepreneurship Development (2nd Revised edition),* Mumbai: Himalaya Publication House

Neelima BN and Swaroop TS 2000 Training women for Entrepreneurship, *Social Welfare* **47** 3–6

Newberry RC 2004 Cannibalism, in *Welfare of the laying hen* (eds) GC Perry, Wallingford: CABI Publishing, 239–258 pages

Nirmala devi M 1997 Developing a strategy for guava production, M.Sc. (Ag.) Thesis, AC&RI, TNAU, Madurai

Nirmala L, Ravichandran V, and Rathakrishnan T 1995 Information sources utilization influencing knowledge and adoption of biofertilizers, *Journal of Extension Education* **6** 1064

Nirmala G and Ravishankar K 2001 Women farm management on rainfed agriculture, *Extension review*, 18–20

Noorjehan AKH 1999 A critical analysis on technological gap in adoption of pest management practice in rice, M.Sc. (Ag.) Thesis, AC&RI, TNAU, Madurai

Nenova T, Niang CT, and Ahmad A 1998 Bringing Finance to Pakistan's Poor: Access to Finance for Small Enterprises, The World Bank Organization for Economic Cooperation and Development, Paris

Oxford English Dictionary, 1993, New Delhi: Oxford University Press

Panda B 1989 Structure and problems of poultry industry in South Asia, *Poultry Guide* **26** 17–26

Pandian S 1999 Video Education: A Tool for knowledge gain by farm women, M.Sc. (Ag.) Thesis, AC&RI, TNAU, Madurai

Pandu Naik G. 2007 Training and Development, New Delhi: Anurag jain for Excel Books.

Pandya RD 1996 Entrepreneurial behaviour of sugarcane growers, *Journal of Extension Education* **6** 1229–1230

Paramasivam A 1988 A critical analysis of the adoption behaviour of paddy seed producers, M.Sc. (Ag.) Thesis, AC&RI, TNAU, Madurai

Parvathi S 1995 A field experiment on farm women's cognitive domain relating to post harvest technologies, M.Sc. (Ag.) Thesis, TNAU, Coimbatore

Gogia P 1999 Developing strategy for coconut production in Andaman and Nicobar Islands, M.Sc. (Ag.) Thesis, AC&RI, TNAU, Madurai

Mathiyalagan P and Subramanian R 1998 Training needs of poultry farmers in scientific poultry management, *Journal of Extension Education* **3** 2099–2101

Perumal G 1990 Project Report on Entrepreneurial characteristics of successful women entrepreneurs - A case analysis approach, Tamil Nadu Agricultural University, Coimbatore, Tamil Nadu

Perumal G and Netaji Seetharaman R 1997 Farm families in diversified needs of small activities, *Tamil Nadu Journal of Extension Education* **1** 127–130

Perumal L 1994 Impact of TANWA training on participants, M.Sc. (Ag.) Thesis, AC&RI, TNAU, Killikulam

Phukan SD 1992 Success of Entrepreneurship Development Programmes, *Yojana* **36** 30–31

Ponmani M 1993 Rural women's participation in the trainings offered by government and voluntary agency, M.Sc. (Ag.) Thesis, AC&RI, TNAU, Madurai

Ponnusamy KA 1990 Training needs of farm women in agriculture and allied fields, *Tamil Nadu Journal of Extension Education* **1** 124–127

Poovannan V 1995 Communication behaviour of TANWA trainees, M.Sc. (Ag.) Thesis, AC&RI, TNAU, Killikulam

Pradeep Kumar R 1993 Aspiration of educated unemployed youth for self-employment in agriculture and allied fields, M.Sc. (Ag.) Thesis, KAU, Thrissur

Prameela K 1992 An ex-post facto study on communication channels utilization behaviour of farm women, M.Sc. (Ag.) Thesis, AC&RI, TNAU, Madurai

Premavathi R 1997 Rural women in farm and home decision making, M.Sc. (Ag.) Thesis, TNAU, Coimbatore

Pushpa J 1992 Training needs of farm women, *Journal of Extension Education* **6** 120–123

Raheem A 2006 Role of SHGs, *Yojana* **50**

Raji RJ, Somasundaram S, and Viswanathan N 1994 Socio-economic profile of trained farmers on irrigation management practices in Thanjavur District, *Journal of Extension Education* **5** 935–936

*Ramakrishnan K 1979 Interview Techniques as a Method of Selection, in *Identification and Selection of Small Scale Entrepreneurs* (eds) TV Rao and TK Moulik, 88–93 pages

Ravichandran M 1997 Participation and Decision making pattern of sericulture farmers, M.Sc. (Ag.) Thesis, AC&RI, TNAU, Madurai

Ravichandran VM 1992 Effectiveness of video education among rural women, M.Sc. (Ag.) Thesis, AC&RI, TNAU, Madurai

Ray MN 1993 Post training evaluation of the training course on extension technique for livestock; Development for North-Eastern Region, *Indian Journal of Social Research* **34** 49–54

Reddy BP 1989 Role of state bank of India in entrepreneurship development in India, in Entrepreneurship Development in India (ed) Samiuddin, New Delhi: Muttrl publishers

Reddy G 2002 Empowering Women through Self Help Groups and Micro Credit; The Case of NIRD Action Research Projects, *Journal of Rural Development*, **21** 511–535

Remi-Alarape AA, Adetayo ED, and Nassar ML 2009 Understanding Entrepreneurial Female Entrepreneurs, *Proceedings of 2nd International Conference on Business Management*

Renuka V 2001 *Opportunities and challenges for women in business, India Together,* Online Report, Civil Society Information Exchange Pvt. Ltd.

Rexlin and T. Selvin Jebaraj Norman 2001 Research Paper on Empowerment of women for HRD, In National seminar on gender issue: Women in agriculture and management, TNAU, Coimbatore.

Rexlin R and Subramanian VS 1990 Women's participation in decision making on farm activities, *Tamil Nadu Journal of Extension Education* **5** 161–165

Rogers EM and Shoemeker FF 1971 *Communication of innovations - A cross cultural approach*, New York: The free press

Robinson S and Stubberud HA 2009 Sources of Advice in Entrepreneurship: Gender Differences in Business Owners' Social Networks, *International Journal of Entrepreneurship*, **13** 83–101

Rodenburg TB and Koene 2004 Feather pecking and feather loss, In *Welfare of the Laying Hen* (ed) GC Perry, Wallingford: CABI Publishing

Roomi MA, Harrison P, and Beaumont-Kerridge J 2009 Women-Owned Small and Medium Enterprises in England: Analysis of Factors Influencing the Growth Process, *Journal of Small Business and Enterprise Development*, **16** 270–288

Roomi MA and Harrison P 2010 Behind the Veil: Women-Only Entrepreneurship Training in Pakistan, *International Journal of Gender and Entrepreneurship* **2** 150–172

Rose S, Camillus L 1991 Training needs of chilli growers, M.Sc. (Ag.) Thesis, AC&RI, TNAU, Madurai

Rosti L and Chelli F 2009 Self-Employment among Italian Female Graduates, *Education and Training* **51** 526–540

Ruth A and Nina H 2005 Measuring empowerment in practise: Structuring analysis and framing indicators, *World Bank policy research working paper*, 3510

Sailaja A and Reddy MN 1999 Training needs of farm women in paddy and additional activities, *Journal of Extension Education* **10** 2420–2424

*Samiuddin 1987 Entrepreneurship Development - A key to Economic Growth, in *Entrepreneurship Development in India* (ed) Samiuddin, New Delhi: Muttrl Publishers

*Sarmah RC and Singh AK 1994 Determinants of Entrepreneurship in Agriculture, *Agriculture Extension Review* **35** 538–539

Savithiri A 1992 A study on knowledge, attitude and adoption of dryland technologies, M.Sc. (Ag) Thesis, AC&RI, TNAU, Madurai

Schumpeter JA 1934 *The Theory of Economic Development,* Cambridge: Harvard University Press

Seethalakshmi R 1999 Designing of training strategies for farm women on integrated pest management in rice, M.Sc. (Ag.) Thesis, AC&RI, TNAU, Madurai

Sehrawat PS, Makhija VK, and Sunag KS 1999 Marketing and Export related problems encountered by entrepreneurs to facilitate setting up of small scale agro processing units, *Indian Journal of Agriculture Marketing* **13** 45–47

Sekar U, Muthiah M, and Annamalai R 1990 Training needs of viticulturists, *Tamil Nadu Journal of Extension Education* **1** 63–70

Sengupta N 1998 Empowerment: A Socio-psychological approach to Self–Help Group Formation, *Prajnan* **26** 525–531

Sethy B 1982 A study on the entrepreneurship characteristics of farmers of an agriculturally advanced district of Orissa, M. Sc. (Ag.) Thesis, IARI, New Delhi

Setty DE 1990 Developing Entrepreneurship in Remote Tribal Communities, *Indian Journal of Social work* **5** 87–89

Shanmugavadivu N 1992 Multi-dimensional role performance of rural women in Farm and Home - An analysis, M.Sc. (Ag.) Thesis, AC&RI, TNAU, Madurai

Sharma DK and Singh TR 1970 Participation of rural women in decision making process related to farm business, *Indian Journal of Extension Education* **6** 43–48

Shah NA 2002 Changing Role of Women in Pakistan: A Study of Social and Economic Activities of Women Entrepreneurs, *Pakistan Research Repository* (available at http://eprints.hec.gov.pk/)

Shane S and Venkataraman S 2000 The Promise of Entrepreneurship as a Field of Research, *Academy of Management Review* **25** 217–226

Singh G and Belwal R 2008 Entrepreneurship and SMEs in Ethiopia: Evaluating the Role, Prospects and Problems faced by Women in this Emergent Sector, *Gender in Management: An International Journal* **23** 120–136

Sorenson RL, Folker CA and Brigham KH 2008 The Collaborative Network Orientation: Achieving Business Success through Collaborative Relationships, *Entrepreneurship: Theory and Practice* **32** 615–634

Sheela 2009 The status of fisherwomen in Andhra Pradesh, *Indian Journal of Gender Studies* **16** 401–410

Sheela T 2000 Information source utilization pattern of farm women in rainfed ecosystem, M.Sc. (Ag.) Thesis, AC&RI, TNAU, Madurai

Singh JN, Tyagi KC, and Sohal TS 1989 Training needs of dairy farmers, *Indian Journal of Extension Education* **15** 77–81

*Singh NP 1979 Tribal Entrepreneurship - Efforts and Achievements, In *Identification and Selection of Small scale Entrepreneurs* (eds) TV Rao and TK Moulik, 160–171 pages

*Singh NP and Sehgal P 1986 Successful Women Entrepreneurs - Their Identity, Expectations and Problems (An Exploratory Research study), NIESBUD, New Delhi

Singh RP 2000 *Management of Training Programmes*, New Delhi: Anmol Publications Pvt. Ltd.

Singh PN 1984 *Training for Management Development*, Mumbai: Suchandra Publications

Sivarevathy AL 1996 Participation of farm women in developmental programmes, M.Sc. (Ag.) Thesis, AC&RI, TNAU, Madurai

Sreen PS and Sreen JS 1992 Constraints Faced by Women Dairy Farmers in India and Strategies toss overcome them, *Indian Journal of Social work* **53** 105–114

Sridaran C 1981 Consequences of Adoption of Sericulture Technology, M.Sc. (Ag.) Thesis, TNAU, Coimbatore

Stevenson H 1983 *A Perspective on Entrepreneurship,* Cambridge, MA: Harvard Business School Press

*Stressman WP 1968 Technological Change and Economic Development - The Manufacturing Experience of Mexico and Puerto Rico, New York: Cornell University Press

Sudan FK 2004.Revitalizing micro-income generating activities through SHG's and micro-credit for rural women in globalization, growth and poverty, New Delhi: Serials Publication

Sudharani K, Sreelatha K, and Surendra G 2000 Empowerment of women in rural areas, *Rural India* 195–198

Sundararajan L 1995 Developing a model for farmers training, M.Sc. (Ag.) Thesis, TNAU, Coimbatore

Supe SV 1969 Factors related to different degrees of rationality in decision making among farmers, Ph.D. Thesis, Division of Agricultural Extension, IARI, New Delhi

Suresh A 1997 Participation in Interface Forestry Programme (IFP) - A Beneficiary Analysis, M.Sc. (Ag.) Thesis, AC&RI, TNAU, Madurai

Suseela MT 1998 Retrospects and Prospects of Apiculture, M.Sc. (Ag.) Thesis, TNAU, Coimbatore

Sutha T 1989 Differential knowledge level and adoption behaviour of Rubber growers, M.Sc. (Ag.) Thesis, AC&RI, TNAU, Madurai

Swaminathan MS 2003 Creating the instruments for a knowledge revolution in rural India, *The Hindu*, July 20

Syamkumar BG 1999 Extension intervention for sustaining rice production, M. Sc. (Ag.) Thesis, KAU, Vellayani

Tambunan T 2009 Women Entrepreneurship in Asian Developing Countries: Their Development and Main Constraints, *Journal of Development and Agricultural Economics* **1** 27–40

Taneja BS 1990 Prospects and problems of poultry farming in Punjab, *Poultry Guide* **27** 81–84

Taori K and Chakravarty p 1994 Rural development through Bee Keeping, *Journal of Rural Development* **13** 565–581

Thamarai SK 1989 Developing a training model for viticulturists, M. Sc. (Ag.) Thesis, AC&RI, TNAU, Madurai

Thangachamy M 1993 Knowledge level and training need of paddy seed growers, M.Sc. (Ag.) Thesis, AC&RI, TNAU, Madurai

Thara J 1996 A study on Kitchen Gardening Behaviour of urban women, M. Sc. (Ag.) Thesis, AC&RI, TNAU, Madurai

Thenamudha V 1996 Entrepreneurial behaviour of farmers - A critical analysis, M.Sc. (Ag.) Thesis, TNAU, Coimbatore

Thiruvirchezhian K 1985 A study on the training needs of scheduled caste and other caste small farmers, M.Sc. (Ag.) Thesis, AC&RI, TNAU, Madurai

*Timmons JA 1986 Growing up big entrepreneurship and creation of high ventura, in *The Art and Science of Entrepreneurship* (eds) LS Donald and WS Ragmon, Cambridge: Ballinger Publishing Company

Timmons JA 2009 New Venture Creation: Entrepreneurship for the 21st Century (8th edition), New York: McGraw-Hill International

Uma K 1994 Vocational settlement pattern of TRYSEM trained rural women, M.Sc. (Ag.) Thesis, AC&RI, TNAU, Madurai

Usha M and Pujar A 1996 Home Science Training at Krishivigyan Kendra, Hanumanamatti, *Agriculture Extension Review* **3** 23–25

Veeraputhiran K 1988 Training needs of fisher women, M.Sc. (Ag.) Thesis, AC&RI, TNAU, Madurai

Velumani 1988 A study on information source utilization and knowledge gap on cotton plant protection technology, M.Sc. (Ag.) Thesis, TNAU, Coimbatore

Venkataramaiah P and Manjula S 1996 Problems faced by Women Entrepreneurs and their suggestions, *Journal of Extension Education* **7** 1451–1453

Venkatesan P 1997 Training needs of rainfed cotton growers, M.Sc. (Ag.) Thesis, AC&RI, TNAU, Madurai

Vidya 2007 Entrepreneurship – Need of the hour (1st edition), Mumbai: Himalaya Publication House

*Vijaya Lakshmi V 1992 Women Entrepreneurship, in *Women's Resource and National Development-A Perspective* (ed) R Chandrasekhar, New Delhi: Gaurav Publishing House

Vijayakhader 1994 Rural women as Entrepreneurs in Mushroom cultivation, *Indian farming* **43** 18–21

Wennekers S and Roy T 1999 Linking entrepreneurship and economic growth, *Small Business Economics* **13** 27–55

Whirter MC 2002 Operationalizing Empowerment in Bank operations, Kennedy School of Governance, Haoard University

Winn J 2005 Women Entrepreneurs: Can we Remove the Barriers? *International Entrepreneurship and Management Journal* **1** 381–397

Wilkins LJ, Brown SN, Zimmerman PH, Leeb C, and Nicol CJ (2004) Investigation of palpation as a method for determining the prevalence of keel and furculum damage in laying hens, *Veterinary Record* **155** 547–551

World Bank 2001 Engendering development: through gender equality in rights, resources, and voice – summary, Washington DC: World Bank

Zamir Ahmed SK 1998 Effectiveness of training to farm women by KVK, M.Sc. (Ag.) Thesis, TNAU, Coimbatore

Zamir Ahmed SK, Philip H, and Venkattakumar R 2000 Preferences of farm women towards training methodology on Agriculture and Allied Fields, *Journal of Extension Education* **11** 2689–2696

* Originals not seen

WEBSITES

info@ijrcm.org.in

www. Smallindustryindia.com

www.dcmsme.gov.in/schemes/Schemes for the development and promotion of women entrepreneurs

www.ghallabhansali.com

www.icfaijournals.com

www.imer.com

www.ludhianadistrict.com

www.newsweek.com

www.udyogini.org

www.zenithresearch.org.in

http://www.gemconsortium.org

http://www.usaid.go.pk

http://www.tameer.org.pk

www.microfinanceconnect.info.

http://www.thethomasyaccatogroup.ca

http://www.cipe.org/

INDEX

Achievement motivation 3
Additional income 56
Agro enterprises 6
Assessment of training needs 66, 67
Characteristics of entrepreneurs 18
Conceptual model 37
Confident 6
Empowerment 24
Entrepreneur 8, 9
Entrepreneurial activities 5
Entrepreneurial awareness 5
Entrepreneurial behaviour 10
Entrepreneurial development programme 3
Entrepreneurship 4, 8
Factors influencing new business 23
Female entrepreneurs 10
Financial assistance 3
Health problems 109, 113
Identification of training needs 67
Impediments 31
Implications 123
Inbuilt quality 2
Initiative 5, 19
Issues 125
Lack of management experience 3
Layer production 98
Managerial capabilities 4
Managerial competence 19
Marketing of mushrooms 88, 115
Medicinal and nutritional value 87
Motivational factors 20
Mushroom production 79, 84
Need for developing entrepreneurship 4
Opportunities 2
Oyster mushroom 86
Positive projection 20
Push and pull factors 1
Reasons for slow progress 2
Rural poverty 4
Self–employment 4, 49
Self–confidence 3
Significance 6, 54
Social factors 21
Social outlook 2
Socio– psychological determinants 12, 55
Stages of empowerment 38
Training need assessment 63, 73
Underemployment 4
Unemployment 4
Upgrading 4
Value addition of fruits and vegetables 7, 90
Vegetable industry 95
Vent pecking 111
Women entrepreneurs 1